Lasers in Manufacturing

IFS

LASERS IN MANUFACTURING

AN INTRODUCTION TO THE TECHNOLOGY

J. T. Luxon
D. E. Parker
and P. D. Plotkowski

IFS (Publications) Ltd, UK
Springer-Verlag
Berlin · Heidelberg · New York
London · Paris · Tokyo

James Luxon
GMI Engineering and Management Institute
1700 West Third Avenue
Flint, MI 48502–2276
USA

British Library Cataloguing in Publication Data
Luxon, James T.
    Lasers in manufacturing
    1. Laser industry
    I. Title   II. Parker, David E.
    III. Plotkowski, Paul D.
    670.42   T5176

ISBN 0–948507–42–X IFS (Publications) Ltd
ISBN 3–540–17427–3 Springer-Verlag Berlin
ISBN 0–387–17427–3 Springer-Verlag New York

© 1987 **IFS (Publications) Ltd,** 35–39 High Street, Kempston,
                                Bedford MK42 7BT, UK
    and **Springer-Verlag**   Berlin Heidelberg New York
                                London Paris Tokyo

Phototypeset by Fleetlines Typesetters, Southend-on-Sea, Essex
Printed by Bartham Press Ltd, Luton

Contents

Preface

THE PURPOSE of this book is to provide an overview of lasers and applications in manufacturing. No attempt has been made to make the book all inclusive. The primary emphasis is on more traditional manufacturing, as opposed to the microelectronics industry, for example.

An attempt has been made by the authors to maintain a minimum of mathematics. Calculus is not required for reading the book, but some knowledge of algebra, trigonometry, and physics would be helpful.

The book is not intended for scientists or engineers experienced in manufacturing applications of lasers. However, it should be helpful to either scientists or engineers just getting into the field of lasers in manufacturing. Engineering, science, and technology students should also find the book useful as an easy reading preview for more advanced study.

The unique properties and the nature of the laser are discussed in the first two chapters. In Chapter Three the types of lasers used in manufacturing applications are described and in Chapter Four those characteristics of applications which make them suitable for the use of lasers are presented. Chapters Five, Six and Seven deal with material processing applications. The principles and applications of laser systems in inspection, measurement and control are discussed in Chapters Eight and Nine. The final chapter deals with laser systems for materials processing applications. Throughout the book references are made to potential or probable future applications.

A bibliography of further reading is given at the end of the book. The chapters to which they apply are indicated by the numbers in bold type and brackets at the end of the references. This list includes useful trade journals and buyers guides which list pertinent information on laser-related equipment manufacturers. Many of the references provide more advanced or more detailed reading on the subjects in this book.

James Luxon
September 1987

Chapter One

Unique properties of laser light

IN THIS chapter the properties of laser light which make lasers such useful tools in an incredible variety of applications are discussed. The way in which these properties relate to manufacturing applications is previewed as an introduction to later chapters. In Chapter Two, the way in which these properties arise from the laser will be explained as well as how the output can be controlled.

The primary properties of concern in industrial applications are radiance (sometimes referred to as brightness, though this is not technically correct), monochromaticity, coherence and the various forms of output, i.e. continuous, pulsed or Q-switched.

Radiance

The radiance of a source of light is the power emitted per unit area of the source per unit solid angle. The units are watts per square metre per steradian. A steradian is the unit of solid angle which is the three-dimensional analogue of a conventional two-dimensional (planar) angle expressed in radians. Hence, the solid angle unit is called a steradian (stereo-radian) as in three-dimensional or whole angle. For small angles the relation between a planar angle and the solid angle of a cone with that planar angle is to a good approximation:

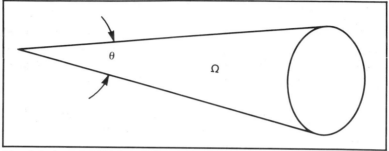

Fig. 1.1 A cone of light of planar angle θ and solid angle Ω

$$\Omega = \frac{\pi}{4}\,\theta^2 \tag{1.1}$$

where θ is the planar angle and Ω is the solid angle. This is illustrated in Fig. 1.1.

The term brightness is a term frequently used instead of the proper term luminance. Luminance is like radiance except that the response of the standard eye is taken into account, and therefore only applies to visible light. The units are lumens per unit area per unit solid angle. A lumen is the photometric equivalent of the radiometric unit (absolute unit of power), the watt. The field of photometry takes into account the eye response whereas radiometry deals with absolute physical quantities. In this book only radiometric quantities will be discussed.

Lasers generally have an extremely high radiance output. For example, a helium-neon (HeNe) laser with an output of 1mW will have an output spot diameter of about 1mm and a beam divergence (angle of the cone into which the light is spreading) of about 1mrad*. Thus, the solid angle:

$$\Omega = \frac{\pi}{4}\,(1\text{mrad})^2 = 0.8 \times 10^{-6}\,\text{sterad}$$

and the radiance:

$$R = \frac{1\text{mW}}{\pi\,(0.5\text{mm})^2\,0.8 \times 10^{-6}\,\text{sterad}} = \frac{160 \times 10^6\text{W}}{\text{m}^2\,\text{sterad}}$$

---

* radian is abbreviated rad

The radiance of the sun is only 1.3×10^6 W/m^2·sterad even though the sun emits a power of 4×10^{26} W.

Radiance is an extremely important concept. High radiance obviously means that the beam has a small divergence angle unless the power is very large. Lasers achieve high radiance at relatively low power levels.

There is a fundamental law of optics which says that the radiance of a source cannot be increased. There are, of course, many ways in which it can be decreased.

It is also true that the amount of power that can be concentrated on a spot by focusing a beam of light is directly proportional to the radiance. In laser work an important concept is the irradiance (frequently called intensity). Irradiance is the power per unit area falling on a surface at a given point. For convenience the units are usually given as watts per square centimetre (W/cm^2). The HeNe laser beam mentioned above could be focused by a 2.5cm focal length lens to an irradiance of over 200W/cm^2. The spot diameter would be only about 25μm.

The high radiance of lasers (small beam divergence) results in small spot sizes when mirrors or lenses are used to focus the beam. This is what then produces the very high irradiance.

In effect, the beam of light from a laser is a narrow cone of light that appears to be coming from a point. As such, geometry generally plays no role in the focusing of a laser beam, at least not the raw beam or one that has simply been expanded or reduced in size. The limiting effect is diffraction. Diffraction causes collimated beams of light to spread (diverge) and causes bending of light beams around sharp edges of objects. It also limits the size of a focused spot. A lens with an f-number* of one will focus an ideal diffraction limited laser beam to a spot radius approximately equal to the wavelength of the light.

---

* f-number is the ratio of lens focal length to aperture or beam size

Clearly, a 1mm focal length lens would be impractical, but if the beam is expanded to 1cm in diameter, then a 1cm focal length lens would be practical and would achieve the desired result. Places where such small spot sizes might be required are integrated circuit manufacturing and recording and reading of optical data storage media such as laser video or audio discs and optical memories for computers. High irradiances are also important to such materials processing applications as cutting, drilling and welding; however, the spot size required is of the order of 0.1–1.0mm for most applications. One thousand watts focused to a 0.1mm diameter spot produces an average irradiance of nearly 13MW/cm^2. This is still very high compared with what can be achieved with conventional sources.

Monochromaticity

The term monochromatic literally means single colour or single wavelength. However, no light source (or any electromagnetic source for that matter) is perfectly monochromatic. Lasers tend to be relatively monochromatic. However, this depends on the type of laser, and special techniques can be used to improve monochromaticity to one part in 10^8 for standard commercial systems for interferometric measurement applications or much better in scien-

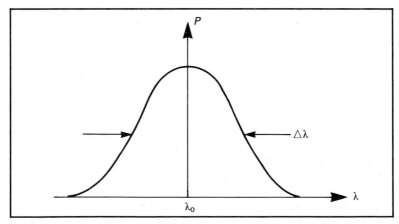

Fig. 1.2 *Power versus wavelength for a fluorescent line*

tific lasers for seismographic work or frequency standardisation.

Lack of perfect monochromaticity is quantitatively expressed in terms of a wavelength bandwidth or frequency bandwidth. Fig. 1.2 illustrates the dependence of power on wavelength. The frequency bandwidth for a typical HeNe laser is about 1500MHz (full width at half-maximum, FWHM). At a wavelength of 0.6328μm this means a wavelength bandwidth of about 0.02Å. The bandwidth of a diode laser with a wavelength of 0.9μm is about 1Å compared with an LED which has a bandwidth of approximately 300Å.

Monochromatic output, or high frequency stability, is of great importance for lasers being used on interferometric measurements since the wavelength is the measure of length or distance and must be known with extreme precision, at least one part in 10^8, and it must remain constant with time. The same holds true for lasers used in chemical and many other scientific analytical applications. Precise wavelength definition is not generally considered important in materials processing though some infrared wavelengths are thought to work better in cutting or drilling of certain polymers, but generally in processes controlled by thermal effects precise wavelength control is not critical. However, processing of polymeric materials with ultraviolet wavelengths, such as are available from excimer lasers, is entirely different. Thermal effects are negligible, molecular bonds are broken directly by interaction with the ultraviolet light. In this case precise tuning of the wavelength to resonate with the molecular bond will probably significantly enhance the efficiency of material removal.

Coherence

The concept of coherence is one of considerable interest in a wide variety of laser applications such as communications, holography, doppler velocimetry and interferometric measurements. Lasers provide a high radiance source of light with a high degree of coherence. This combination is not available from any other source of light.

Light is a part of the electromagnetic spectrum (0.4–0.7μm) and as such it can be thought of as a sinusoidal travelling wave represented by:

$$E = E_0 \sin(2\pi ft - \frac{2\pi}{\lambda} x) \sin(2\pi ft \qquad (1.2)$$

where E is the electric field intensity in volts per metre, E_0 is the electric field intensity amplitude, f is frequency in hertz and λ is the wavelength in metres. Equation (1.2) represents an infinite plane wave (infinite extent is the y–z plane) which is perfectly monochromatic and has perfect coherence. This is true because it is a pure sine wave extending from $x = -\infty$ to $x = +\infty$. Real light waves are more accurately represented by a sum of an infinite number of waves of infinitesimally differing frequencies (wavelengths). This leads to waves that are localised in space (called wavepackets or photons) that have a certain frequency (wavelength), bandwidth and less than perfect coherence.

To illustrate what is meant by coherence, consider just two waves of slightly different frequencies which are added together:

$$E_1 = E_0 \sin(\omega t - \kappa x)$$
$$E_2 = E_0 \sin[(\omega + \triangle\omega)t - (\kappa + \triangle\kappa)x] \qquad (1.3)$$

where $\omega = 2\pi f$, $\kappa = 2\pi/\lambda$ and $\triangle\omega$ and $\triangle\kappa$ are small. When E_1 and E_2 are added, the result is:

$$E = 2E_0 \cos(\omega t - \kappa x) \cos\left(\frac{\triangle\omega t}{2} - \frac{\triangle\kappa x}{2}\right) \qquad (1.4)$$

Fig. 1.3 is a sketch of E vs x for a specific time. The 'packets' or envelope is moving at the speed $v_g = \triangle\omega/\triangle\kappa$, called the group velocity. The amplitude variation given by the $\cos\left(\frac{\triangle\omega t}{2} - \frac{\triangle\kappa x}{2}\right)$ term is caused by the periodic variation from constructive to destructive interference of the two original waves. The length of one of the 'beats', as they are called (this is analagous to the beating effect of the sound from two aircraft or boat engines which are not quite synchro-

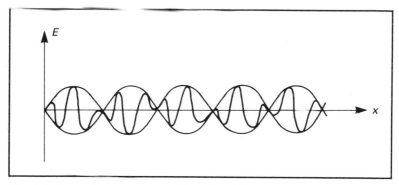

Fig. 1.3 Sum of two waves of slightly different frequencies

nised), is $2\pi/\triangle\kappa$. Roughly, one could say this is the distance over which the two interfere constructively or are 'in phase', before they become out of phase. The waves are coherent in this distance and we can refer to L as a coherence length. The coherence time is L/v_g which for light in free space is L/C, where $C = 3 \times 10^8$m/s. Notice that small $\triangle\kappa$ (small $\triangle\omega$) implies large L.

The foregoing discussion only applies to the addition of two waves, but the conclusions are essentially valid for light in general when an infinite number of waves are summed, only one pulse (a photon) occurs and the relationship between $\triangle\kappa$ and L is approximately:

$$\triangle\kappa L \geqslant 1 \tag{1.5}$$

where $2\triangle\kappa$ gives the FWHM bandwidth. Since $\triangle f = C\triangle\kappa$, Eqn. (1.5) says that to achieve long coherence length a narrow bandwidth is required. For example, a coherence length of 3m will require a frequency bandwidth of $\triangle f = C/2L = 500$MHz. For a HeNe laser the central frequency is 4.74×10^{14}Hz so the required bandwidth is $1 \times 10^{-5}\%$ of the central frequency – a narrow bandwidth indeed. This coherence length is typical of HeNe lasers used in holography whereas the coherence length for a HeNe laser used in an interferometric displacement measurement instrument may be 100m, which requires a frequency bandwidth and stability of better than one part in 10^8.

In reality there are two types of coherence; temporal, which is what we have been talking about, and spatial. Temporal coherence refers to correlation in phase at a given point in space over a length of time. Coherence time $\tau = L/C$, is the time it takes a packet (photon) to pass a given point. Spatial coherence refers to correlation in phase at different points at the same time. Of particular importance is the phase correlation on a surface transverse to the direction of propagation called a phase front. For a perfectly monochromatic wave this would be called a wavefront (e.g. the crest of a waterwave) and all points would have the same phase. For lasers, phase fronts are spherical and the spatial coherence for TEM_{00} lasers is excellent.

Good temporal coherence is essential for interferometric applications such as holography. Good transverse spatial coherence is required for certain types of diffraction applications. High transverse spatial coherence causes the characteristic laser 'speckle' which is used for interferometric strain measurements, a technique called speckle interferometry. High spatial coherence causes undesirable diffraction effects when laser light is used to illuminate small details such as integrated circuit masks. Fortunately, not all lasers have high temporal and spatial coherence. Excimer lasers, which are used in integrated circuit manufacturing, represent a case in point.

To summarise the discussion of coherence, a marching band analogy for laser output is suggested. In a parade of marching bands one would see nearly perfect spatial coherence across the rows (phasefronts) and a band length (coherence length) of 10–20 rows (average perhaps 15). It would not be possible to predict the step (phase) of one band bsaed on observations of another, yet they are all marching at the same speed (speed of propagation). An incoherent light source is more nearly analogous to the people walking on a busy city sidewalk at rush hour.

Power and energy output

The output of industrial lasers may be pulsed or continuous. Lasers used for alignment, measurements and other low-power applications usually have a continuous output whereas materials processing lasers may be either pulsed or continuous. Some lasers are capable of both pulsed and continuous operation.

Continuous output is referred to as CW (continuous wave) output. Literally, however, it simply means that the power output is constant. Industrial CW lasers must be capable of maintaining constant power output within narrow limits over a long period of time. A typical specification for a high-power CO_2 laser is a $\pm 2\%$ long-term variation in power – long-term meaning at least 24 hours. The power output is given in watts (W) for CW lasers and the total energy output in joules in a given length of time is simply power $\times$ time or Pt.

Pulsed output may be accomplished by a variety of techniques. Material processing lasers are usually pulsed electronically. In gas lasers, capacitive or inductive power supplies capable of producing high-power (large energy) pulses are used. Pulse shaping networks are employed to produce the most advantageous pulse shape, such as a 5ms pulse with a sharp leading edge spike to break down the initially high reflectance of a metal being welded. Peak powers for pulsed CO_2 lasers may exceed the average power by a factor of 3 to 4 for capacitive power supplies and by as much as 10 for rf or inductively pulsed systems. Pulse repetition rates of several thousand hertz can be achieved.

Solid lasers such as Nd-Glass and Nd-YAG are also electronically pulsed with flashlamps to achieve high peak power. The output power of a pulsed solid laser is not a smooth function of time. Fig. 1.4 shows sketches of pulses from Nd-Glass and Nd-YAG lasers. The pulses are actually a series of spikes of irregular height and spacing. The Nd-YAG laser tends to relax to a smooth output towards the end of the pulse. The highly spiky character is most desirable for drilling, although proper optimisation of drilling parameters permits high-quality drilling with Nd-YAG lasers. Pulse shaping can be used to alter the temporal distribution of energy in the pulse. Pulse lengths for power supply pulsing are typically 0.1–10ms depending on the application.

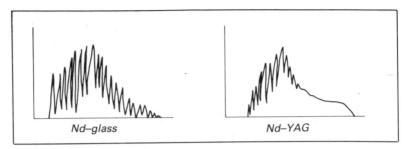

Fig. 1.4 Output pulses for solid lasers

A process referred to as Q-switching is used to produce pulses in the ns to μs range. Q-switching is used to produce short double or triple pulses for pulsed holography with ruby lasers and is also used with Nd-YAG marking and resistor trimming lasers. In the latter case, a thousand-fold increase in peak power over average power can be achieved with repetition rates of several thousand pulses per second.

Chapter Two

What is the laser?

THE LASER is in all cases an extremely bright or high radiance source of light compared with other light sources of comparable power output. The laser can also be highly coherent and monochromatic, if this is required. The high radiance leads to the ability to focus laser beams to spot sizes of the order of the wavelength of the light.

The acronym LASER stands for Light Amplification by Stimulated Emission of Radiation. The word *light* refers to ultraviolet, visible or infrared electromagnetic radiation. The word radiation refers to *electromagnetic radiation*, of which light is a special case. The key words in the acronym are *amplification by stimulated emission*. A laser amplifies light and it does this by means of the phenomenon of stimulated emission. Most, but not all, lasers are also oscillators because an optical resonator is used to provide feedback and energy storage.

In the remainder of this chapter some key concepts are discussed to provide an understanding of how the laser works and why it produces the unique properties of light that it exhibits. The nature of the beam, how it propagates, and the effects of focusing are discussed. The general classes of lasers are also described; specific types of lasers are discussed in Chapter Three.

Laser concepts

Physical nature of the laser

The industrial lasers ~~that will be discussed in Chapter Three~~ are all oscillators and as such an optical resonator is an integral part of each type. In its basic form, laser resonators consist of two spherical or plane mirrors aligned parallel and facing one another on an axis, as depicted in Fig. 2.1. The back mirror is close to being 100% reflective. In some cases a small amount of power, less than 1%, is allowed to pass through the back mirror to a power meter, the output of which may be used to stabilise the output of the laser automatically. The output mirror, also called an output coupler, is partially transparent. As low as 1% or as high as 50% of the power incident on the output mirror is transmitted, depending on the type and power level of the laser.

The industrial lasers discussed in this text are either solid or gas lasers. The only important liquid laser is the continuous wavelength-tunable dye laser which has found extensive scientific applications, but is of too low power output to be of great use in manufacturing.

A solid laser can be schematically depicted by placing a solid rod, usually glass or crystal, between the mirrors with one or two high intensity krypton (Kr) or xenon (Xe) gas discharge lamps placed parallel to the rod (see Fig. 2.2).

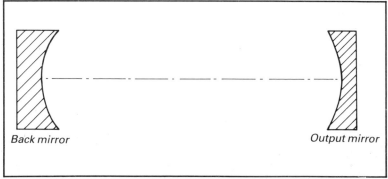

Back mirror   Output mirror

Fig. 2.1 Simple linear optical resonator

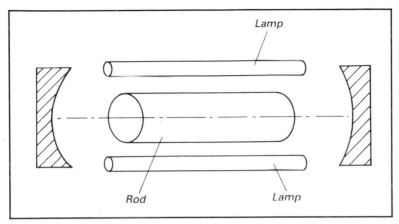

Fig. 2.2 Simple schematic of a solid laser

The schematic in Fig. 2.2 does not show the great many details that go into the construction of a laser, such as the electrical power supply, water cooling, intricate details of the optics, or other intracavity devices that might be included.

For a gas laser the rod and lamps in Fig. 2.2 are replaced, in some cases, by a gas discharge tube which may be air or liquid cooled, depending on the type of laser. A dc or rf voltage is placed across electrodes attached to the tube, much the same as any gas discharge lamp, e.g. a fluorescent light tube.

Population inversion

One phenomenon which must occur for a laser to function, i.e. to *lase*, is a population inversion. Consequently, a qualitative understanding of this phenomenon is essential to understanding how the laser works.

The light that is emitted by a laser is the consequence of the emission of a photon when a transition between an upper and a lower energy level takes place. This may be electronic (electron transition in an atom or ion), vibrational (transition beteen vibrational states of a molecule) or vibronic (a combination of the two). Two such energy levels, E_1 and E_2, are depicted in Fig. 2.3. The little circles represent atoms or

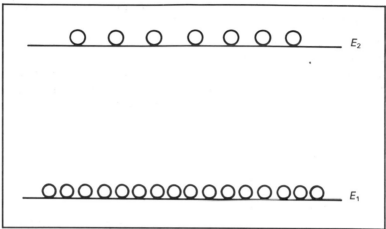

Fig. 2.3 Energy levels for a laser transition

molecules in each state. At *equilibrium* there can never be, on the average, more atoms or molecules with energy E_2 than with energy E_1. In lasers, a *non-equilibrium* situation is created by means of the light from gas-discharge lamps for solid lasers or the electrical discharge for gas lasers. Under certain conditions this non-equilibrium situation creates a population inversion between the states E_2 and E_1. This means that more atoms or molecules are in state E_2 than E_1. This condition is *necessary*, though not sufficient generally, for lasing action to occur. In fact, a certain threshold population inversion must be achieved before lasing occurs. If n_2 is the number of atoms or molecules with energy E_2 and n_1 is the number with energy E_1, then $n_2 - n_1$ must exceed some threshold value for lasing to occur. Anytime the population inversion ($n_2 - n_1$) drops below threshold, lasing ceases.

Optical processes

In this section, the ways in which photons can be absorbed or emitted by a material are discussed. When a photon enters a material, it can be absorbed by exciting an electron to a higher energy level or in the case of striking a molecule it can increase the molecular vibration energy. In crystals, the photon may lose all or a portion of its energy through interactions with crystal vibrations called phonons*. Similar effects occur in other solids and liquids.

Photons may be emitted as a result of spontaneous transitions from electronic or molecular excited energy states to lower energy states or by stimulated emission between such states. Not all transitions involve photon emission – so-called non-radiative transitions occur through interactions with the surrounding medium such as phonon interactions in crystals.

Fig. 2.4 illustrates the optical processes of absorption, spontaneous emission and stimulated emission for two energy levels E_2 and E_1. If the energy of a photon hf is less than $E_2 - E_1$, it cannot be absorbed creating a transition from E_1 to E_2. Likewise if $hf \gg E_2 - E_1$, absorption is highly improbable. Essentially, $hf = E_2 - E_1$ for absorption to occur.

When an atom or molecule is put in an excited state such as E_2, it will have a certain average lifetime τ in that state before undergoing an allowed transition. A typical lifetime is 10^{-8}s, but in lasers the state must be considerably longer for a population inversion to be established. In this case, τ is closer to 10^{-3}s. Particles (generic for atoms or molecules) will be leaving state E_2 by spontaneous emission at the rate of n_2/τ, so if a population inversion is to be established the 'pump' rate must exceed n_2/τ.

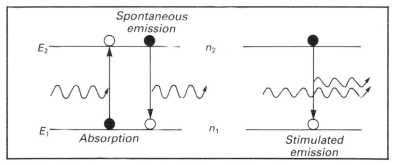

Fig. 2.4 Optical process for transitions between two energy levels

---

* This is the name given to quasiparticles which represent crystal lattice vibrations. These lattice vibrations behave like real particles in certain situations

Stimulated emission is a process whereby a photon passing in close proximity to an excited particle causes it to undergo a transition prior to when it would have spontaneously. The stimulating photon must have an energy $hf = E_2 - E_1$ for this to occur. There is no transfer of energy between the stimulating photon and the particle. However, the excited particle is analogous to a classical oscillating dipole and the photon electromagnetic field interacts with the oscillating dipole basically 'forcing' it to undergo the transition.

Einstein proved that the probability per particle per unit time for stimulated emission is the same as for absorption, and that the rate of spontaneous emission n_2/τ is small compared with stimulated emission in the presence of a moderate density of electromagnetic radiation. This means that if there are enough photons of the right energy (wavelength) hanging around, there will be a lot more stimulated emission than absorption if there is also a population inversion so that these loitering photons are not absorbed before they can stimulate the emission of additional photons.

It is clear that the stimulated photons have the same energy, and therfore same wavelength, as the stimulators. What is not obvious is that the stimulatees and stimulators end up travelling in the same direction and are in phase with one another. This is somewhat analogous to a parade that grows while it progresses as the result of people joining it from the curbside – assuming they get in step of course.

If some mechanism is provided to keep the radiation (parade) moving in a straight line, then you have a laser (maybe). In the parade analogy, anyone who wanders off the parade path stumbles over the curb and is lost to the parade. In a laser, the stray radiation may hit the walls of a tube, rod surface or an aperture in the system and thus is absorbed.

Gain

It was pointed out that lasers are amplifiers. This becomes clear if the effect of stimulated emission in an inverted population is considered. The entire process is started by a

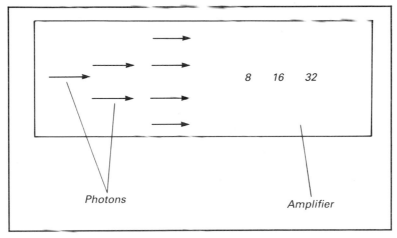

Fig. 2.5 Laser amplifier

spontaneously emitted photon which produces another photon by stimulated emission. These two in turn produce four and these four produce eight, and so on. This is illustrated in Fig. 2.5 which depicts an amplifier and the onset of lasing action. The speed of light is 3×10^8m/s so a photon travels 3m in 10^{-8}s. This means that one photon can produce a large number of new photons through stimulated emission in the 1ms lifetime for spontaneous emission. If the gain is not sufficient in a single pass through the amplifier to produce lasing action (and it usually isn't), mirrors are added to either end to provide a longer effective path. Each photon may, on average, pass through the gain medium several times. For 1% overall reflectance, the average number of photon passes is 100; for 50% overall reflectance the average number of passes is 2. The feedback also helps to increase the linearity or unidirectionality of the output beam.

Excitation methods

The common methods for exciting lasers have already been mentioned. In gas lasers, dc and rf discharges are used. Helium-neon lasers are dc-excited. Many CO_2 lasers are dc-excited, although rf is common particularly for sealed-off

CO_2 lasers such as waveguide CO_2 lasers. Combinations of dc and rf may also be utilised.

The lamps used in solid lasers are Xe or Kr gas discharge lamps similar to the one shown in Fig. 2.6. Xe is somewhat better for pulsed operation, whereas Kr is preferred for CW operation. The lamps are sealed with O-rings in a quartz tube and are surrounded by de-ionised water coolant.

Some dye lasers are excited by other lasers, such as frequency doubled or tripled* Nd-YAG, argon ion lasers or N_2 lasers, which emit in the uv. Other methods that are used are chemical reactions, gas dynamic and even light from the sun. In a hydrogen-fluoride laser, hydrogen and fluorine gases are held in separate tanks and mixed just prior to injection into the active region of the laser. The exothermic reaction which forms HF produces the excitation for the hydrogen-fluoride molecule.

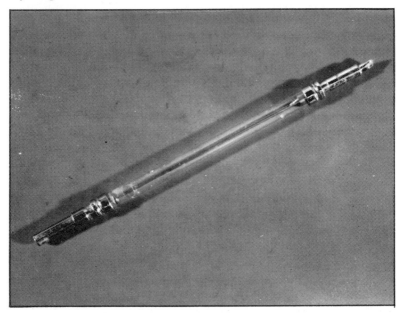

Fig. 2.6 Gas discharge lamp for solid laser

* Frequency multiplying refers to harmonic generation. When an intense laser beam is passed through certain materials, non-linear effects can cause the frequency to be multiplied

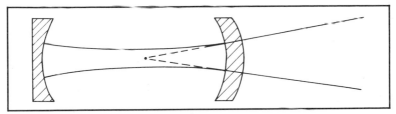

Fig. 2.7 Laser beam (not to scale)

In gas-dynamic lasers a molecular gas such as CO_2 is stagnated at high pressure and temperature and then forced through supersonic nozzles. Upon rapid expansion and cooling, the lower energy levels are depopulated quickly leaving a highly inverted population between excited states. If this is allowed to occur in an optical resonator, lasing action takes place. Very high power levels can be generated in this manner. At present, no commercial systems use chemical or gas-dynamic excitation.

Optics for lasers

Diffraction effects

It was pointed out in Chapter One that even very low-power lasers, 1mW or less, produce radiances which exceed that of the sun. This is chiefly due to the fact that the divergence angle (spreading angle) of a laser beam is usually very small, being of the order of 1–10mrad. Fig. 2.7 is a sketch of a laser beam both inside and outside of a stable* resonator. The lines drawn in Fig. 2.7 are not rays, but instead represent loci of some definition of spot size. The mirrors have spherical surfaces and the laser beam phase fronts† are spherical and conform to the shape of the mirrors when they reach the mirrors. The output mirror will act as a lens unless the inner and outer curvatures are the same. A flat mirror avoids this problem. The output mirror can be used to reduce geometrical divergence by making it into a positive

---

* Stable here refers to the fact that some rays retrace themselves in the resonator as opposed to unstable where this does not occur

† A phase front is an imaginary surface on which the relative phases stay the same as the front moves through space. It would be called a wavefront if all points had the same phase

lens by putting a slightly smaller curvature on the outer surface than on the inner surface. Since the beam has a spherical phase front, the beam always appears to be coming from a point. According to geometrical optics, a point source of light is focused to a point. Unfortunately, the wave properties of light do not permit this to happen. Because light has wavelike properties, diffraction effects play a major role in the propagation and focusing of laser beams. Diffraction is the result of the ability of waves to interfere with one another to produce constructive and destructive interference, as well as shades in between. Diffraction occurs whenever a beam of light is obstructed by an object or aperture or otherwise limited in size. We normally do not observe diffraction effects because ordinary light is incoherent and equal amounts of constructive and destructive interference occur in such a way as to produce essentially uniform illumination. However, diffraction effects can be observed even with ordinary sources such as incandescent lamps or sodium arc lamps if they are far enough away from the diffraction element. This is readily proven by forming a small slit between your fingers and viewing a distant light source (a couple of feet will do) through the slit. The vertical fringes you see are due to diffraction caused by the light coming through the slit.

In a sense, the same thing happens to any source of light with a finite lateral size propagating through space. If you were to image the fringes you saw beween your fingers on a screen or sheet of paper, you would notice that the farther the screen is from your fingers the farther apart are the fringes. You would see a central bright fringe which can be thought of as an image of the slit. The image gets bigger as the screen is moved away because the light is diverging. The divergence angle of the central fringe is approximately:

$$\theta = \frac{\lambda}{D} \tag{2.1}$$

where λ is the wavelength of light and D is the slit width.

Fig. 2.8 illustrates the diffraction phenomenon for a slit in a screen; θ is the half-angle of divergence. Whether the slit is

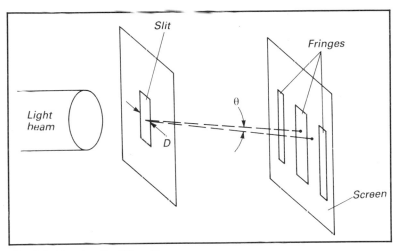

Fig. 2.8 Single slit diffraction

there or not is irrelevant. A beam of monochromatic light of width D will diffract (spread) at approximately the angle θ. A circular aperture will cause a diffraction angle of:

$$\theta = 1.22\frac{\lambda}{D} \tag{2.2}$$

where D is the aperture diameter. Equation (2.2) is a better approximation for circular cross-section monochromatic beams.

The point of all this is that even if a laser is designed, or the beam is modified, so that there is no geometrical spreading, diffraction will cause the beam to diverge because of its finite size and the diffraction angle will *always be inversely proportional to the beam size*. Ordinary light beams exhibit very large diffraction spreading because of the lack of coherence. The beam in this case can actually be thought of as a very large number of parallel independent beams of much smaller size than the overall beam, hence the effective value of D in Eqn. (2.2) is very small.

Thus, laser beams diverge much more slowly than conventional beams and are therefore much brighter (have higher radiance).

If we used a lens to focus the slit in Fig. 2.8 onto a screen placed in the lens focal plane, we would find that the width of the slit image is:

$$d = \frac{2f\lambda}{D} = 2f\theta \tag{2.3}$$

Note that θ here has *nothing* to do with the inherent divergence of the beam illuminating the slit, and if it had any we have ignored it. For a circular aperture the image diameter would be:

$$d = 2(1.22\frac{\lambda}{D})f = 2.44\frac{f\lambda}{D} \tag{2.4}$$

In any case we get a finite size spot or l   mage whose dimensions depend on the wavelength of the light, the slit or aperture size and the lens focal length. For a purely monochromatic wave it should be possible to focus the beam to a spot whose radius is approximately λ if we use a lens whose focal length is approximately the diameter of the beam (this would be an f-1.0 lens, effectively).

Laser resonators

Stable laser resonators produce a variety of output modes as might be expected from an oscillator. The condition that leads to these modes is the fact that in one round trip through the laser the phase front must faithfully reproduce itself. This means that an integral number of half-wavelengths must occur between the mirrors as shown in Fig. 2.9.

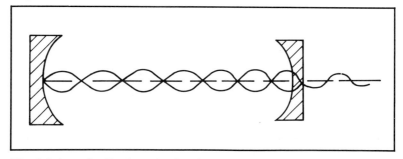

Fig. 2.9 Longitudinal modes in a laser

The condition shown in Fig. 2.9 can be expressed by:

$$n \frac{\lambda}{2} = L \tag{2.5}$$

where n is an integer, and using $c = f\lambda$ where c is the speed of light:

$$f = n \frac{c}{2L} \tag{2.6}$$

The frequency spacing produced by the longitudinal modes is, therefore, $\frac{c}{2L}$.

There are a variety of ways that the phase of the wave can be distributed transversely and this leads to the so-called transverse electromagnetic modes denoted by TEM_{pq} where p and q represent the number of nodes in orthogonal directions, i.e. the p nodes are perpendicular to the q nodes. Fig. 2.10 illustrates some of the possible transverse modes. The $\text{TEM}_{01}{}^{*}$ (doughnut) mode is the result of a TEM_{01} or TEM_{10} where the mode direction is changing rapidly with time. The TEM modes generally result in resonant frequencies lying between the pure longitudinal mode frequencies. The TEM_{00} mode is the most desirable mode for most applications and produces a Gaussian irradiance distribution (power per unit area versus radial distance) as shown in Fig. 2.11.

The irradiance for the TEM_{00} mode is given by:

$$I = I_0 e^{-2r^2/w^2} \tag{2.7}$$

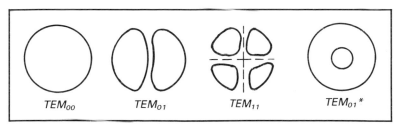

TEM$_{00}$ TEM$_{01}$ TEM$_{11}$ TEM$_{01}$*

Fig. 2.10 Transverse electromagnetic modes

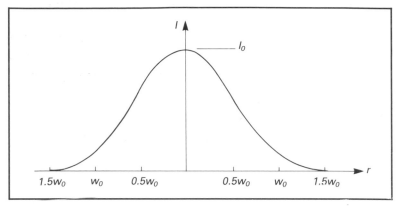

Fig. 2.11 Irradiance distribution for TEM$_{00}$ mode

where I_o is the irradiance at the beam centre and w is the spot radius such that at $r = w$, $I = I_o/e^2 = 0.135\ I_o$. The lines in Fig. 2.7 are loci of points corresponding to the value of w as a function of distance along the beam axis.

Because the TEM$_{00}$ (Gaussian) mode has uniphase phase fronts with a smooth drop off in irradiance, the diffraction spreading is minimum for a laser beam and considerably smaller than given by Eqn. (2.2). Higher order mode beams suffer greater diffraction spreading due to the fact that they are, in effect, composed of two or more individual beams of different phases, hence the effective aperture is smaller leading to greater diffraction spreading.

In gas lasers, proper laser resonator design can take advantage of the more rapid diffraction spreading of the higher order modes so that the output is predominantly Gaussian. One way this is done is to keep the Fresnel number, $N = \dfrac{a^2}{\lambda L}$, close to unity. In this formula, a is the radius of the output aperture and L is the resonator optical path length. N is the number of diffraction rings that would be observed at the output mirror if the back mirror is uniformly illuminated. Thus, a low Fresnel number means large diffraction losses for the higher order modes. Another technique, frequently applied in solid lasers, is to place an aperture at the waist which is just large enough to let the

TEM$_{oo}$ mode through, but which significantly attenuates the larger higher order modes.

The terms near-field and far-field are frequently used in discussion of laser beams. In the near-field the divergence angle of the beam is steadily, though slowly, increasing. In the far-field the divergence angle is constant. In fact, if the beam radius lines are extended straight backwards from the far-field in Fig. 2.7 they will intersect at the waist. The waist is the narrowest part of the beam. In the far-field $(\frac{\lambda z}{\pi w_0^2})^2 \gg 1$,

where w_0 is the beam radius at the waist. Thus $z_{ff} \gg \frac{\pi w_0^2}{\lambda}$,

which for a HeNe laser would be about 50m. The divergence angle in the far-field is (this is half-angle divergence):

$$\theta_{ff} = \frac{\lambda}{\pi w_0} \tag{2.8}$$

assuming no lensing effect from the output mirror. Again, for a HeNe laser this is about 0.5mrad.

Focusing and depth of focus

When a laser beam is focused by a lens or mirror, the focused spot size can be calculated accurately if the output mode is single mode. For multimode or distorted output, focused spot size can in practice only be estimated on the basis of diffraction by an aperture equal to the beam size. In this case it is best to use Eqn. (2.2). The focused spot size for a Gaussian beam is given to a good approximation by:

$$w_f = \frac{\lambda f}{\pi w_1} \tag{2.9}$$

where w_1 is the beam radius at the lens. It should be recognised that the actual area affected by a focused laser beam may differ substantially from that given by the radius in Eqn. (2.9).

The depth of focus of a focused beam is the allowable distance from the plane of best focus which still produces a spot of acceptable size. This will vary with the application. For a Gaussian beam the depth of focus is given by:

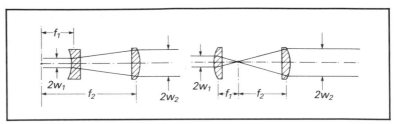

Fig. 2.12 Beam expanders

$$d = \pm \pi \sqrt{\rho^2 - 1} \left| \frac{w_0^2}{\lambda} \right.$$

where w_o is the focused spot size and ρ is a factor such that ρw_o gives the acceptable spot size. For example, if $\rho = 1.05$, then $d = \pm w_0^2/\lambda$. This is the depth of focus for a 5% allowable spot size increase.

Frequently, laser beams are expanded. This may serve one or all three of the following purposes: to decrease divergence, to decrease irradiance, and/or to produce a larger illuminated area. Fig. 2.12 illustrates two commonly used types of transmissive beam expanders; expanders may also be reflective. In either case:

$$w_2 = \frac{f_2}{f_1} w_1 \tag{2.11}$$

and $\theta_2 = \frac{f_1}{f_2} \theta_1$ (2.12)

where θ_1 and θ_2 are the initial and final divergence angles.

Unstable resonators

A few comments should be made about unstable resonators since they are commonly used with high-power industrial lasers. An unstable resonator is one in which there are no non-trivial rays which retrace themselves, i.e. there is no focusing effect by the mirrors. A fairly common unstable resonator design used in high-power lasers is the confocal arrangement depicted in Fig. 2.13.

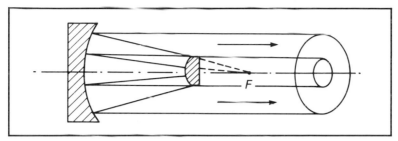

Fig. 2.13 Confocal unstable resonators

Light generated in the resonator geometrically walks off the axis and comes out in the shape of a ring with some diffraction-induced substructure. Such resonators are particularly useful where relatively low reflectance from the output coupler is required, e.g. around 50%. The beam is intercepted by a third mirror and reflected out of the cavity to avoid having the beam obstructed by the output mirror mounts.

When such a beam is focused, the hole in the middle disappears and the irradiance distribution becomes Gaussian-like. Also, the same thing happens in the far-field. Actually, the image in the focal plane of a lens is a far-field pattern. Stable resonator modes *do not* change form between the near- and far-fields.

The main advantage of unstable resonators is that it eliminates the transmissive output mirror. Some designs have no transmissive optics at all in the laser head. The beam exits through an acrowindow which is designed to allow the beam to pass through while laser and atmospheric gasses are mixed in a chamber and pumped out with negligible contamination of the active gas region by atmospheric gasses.

General classifications of lasers

This section is devoted to classifying lasers in broad terms and identifying which industrial lasers fall into these categories. The three major categories refer to the state of matter of the active material, i.e. solid, liquid or gas phase.

Lasers which employ an active medium in the solid state are ruby, which is actually chromium doped crystalline alumina (Cr^{3+} − Al_2O_3), neodymium-yttrium-aluminium-garnet (Nd^{3+}−YAG), Nd^{3+}−glass and the diode laser. The only laser which uses a liquid medium as the active medium is the dye laser which will be found in certain industrial research laboratories. The number of lasers which utilise the gas phase for the active medium is extensive; however, only the CO_2, HeNe and Ar laser have much significance in manufacturing.

With the exception of the diode laser the transitions in solid lasers are electronic transitions of the trivalent ions Cr^{3+} or Nd^{3+} and excitation is by means of high-intensity Xe or Kr lamps. The diode laser operates by recombination of electron-hole pairs in a semiconductor diode as a result of charge injection into the depletion region separating the p- and n-type semiconductor materials.

In dye lasers the transitions take place between hundreds of possible energy levels that exist in dye molecules. These lasers can be tuned continuously, in some cases, from $0.38\mu m$ to $0.75\mu m$ and may be excited by another laser (the most common method) or high-intensity lamps.

Gas lasers may be neutral or ionic lasers such as HeNe which is a neutral gas laser, or Ar which is an ion laser. Ne is the lasant in HeNe lasers. The light that is emitted involves fluorescent lines of neutral Ne. On the other hand, in an ion laser such as Ar, the fluorescence is derived from electronic energy levels of singly ionised argon.

The CO_2 laser is referred to as a molecular laser because the fluorescence is the result of transitions between various vibrational-rotational energy levels of the CO_2 molecule.

One important laser, particularly in the semiconductor industry and as a pump laser for dye lasers, the N_2 laser, is placed in the category of vibronic lasers. Vibronic (vibration and electronic) implies that transitions take place between vibrational levels of different electronic levels. Emission is in the ultraviolet-visible portion of the spectrum.

A very unusual class of lasers which are finding applications in medicine and the semiconductor *manufacturing* industry, and will probably become significant in general manufacturing are called excimer lasers. Strictly speaking, excimer stands for excited dimer, where dimer is the name given to any molecule of the form X_2. An excimer is denoted by X_2^*. Some atoms do not form stable dimers in the ground state, but do in the excited state. When the dimer drops to the ground state it dissociates. Excimer lasers actually utilise an excited molecule of a rare gas atom and a halide. Strictly, these are not excimers because a dimer is not involved, but this is common terminology and the behaviour is essentially the same. The fluorescence produced when the excimer drops to the ground state and dissociates, gives rise to the laser radiation. The excitation energy is very high so these lasers emit in the ultraviolet and are somewhat tunable. Examples are argon-fluoride (ArF), krypton-fluoride (KrF), xenon-fluoride (XeF) and xenon-chloride (XeCl) with a buffer gas of He or Ar. These lasers are excited by an electron beam or an electrical discharge.

The final type of laser to be mentioned here, which really does not fit any of the previous categories, is the free electon laser. Such lasers hold great promise for materials processing because they have wide wavelength tuning capability and potentially very high power levels. High-energy electrons from a linear accelerator are passed through a series of magnets with alternating polarity, called a 'wiggler'. The alternating acceleration imparted to the electrons results in the emission of radiation. Stimulated emission occurs and a free-electron laser (FEL) results. Such lasers can operate from the ultraviolet into the infrared portion of the spectrum.

Chapter Three

Lasers for industry

THIS chapter describes the lasers commonly used in manufacturing and their general areas of application. More detail on these applications will be presented in later chapters, as will possible future developments in industrial lasers.

The lasers to be described at this point are the helium-neon (HeNe), diode, neodymium-ytrium-aluminium-garnet (Nd-YAG), neodymium-glass (Nd-glass), carbon dioxide (CO_2) lasers and excimer lasers.

Helium-neon lasers

The HeNe laser is a neutral atom gas laser. Excitation is by means of a dc glow discharge. There are a number of wavelengths available for HeNe lasers including 3.39m, 1.15m in the infrared and the visible outputs at 0.543μm (green), 0.594μm (yellow), 0.612μm (orange) and 0.6328μm (red). The red wavelength is the most commonly used. HeNe lasers which are tunable over several wavelengths are available.

The maximum power output of commercial HeNe lasers is 50mW for a tube about 2m in length with an efficiency of less than 0.1%. The bore of the tube is kept small (1–6mm) to control the transverse output mode, which is usually Gaussian (TEM_{00}).

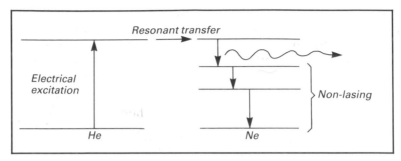

Fig. 3.1   Energy level diagram of HeNe laser (not to scale)

Ne is the lasant in the HeNe laser, but is assisted in an interesting manner by the He. He atoms excited by collisions with electrons in the gas discharge collide with unexcited Ne atoms and transfer their energy to a coincident Ne energy level. The Ne atom then emits a photon when it drops to a lower intermediate energy level. Subsequent transitions return the Ne atom to the ground state so that it can participate in the process all over again. This process is schematically illustrated in Fig. 3.1.

The HeNe laser (Fig. 3.2) has a frequency bandwidth of

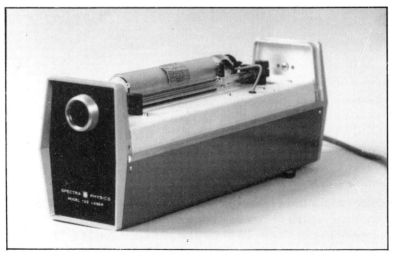

Fig. 3.2   A low-power HeNe laser

about 1500–1700 MHz. Single axial mode operation can be guaranteed by making the laser resonator short (about 10–15cm). Further frequency stabilisation is achieved by means of a servo-controlled mirror attached to a piezoelectric crystal which expands and contracts in response to an applied voltage to keep the cavity length fixed.

HeNe lasers are low-gain lasers, hence mirrors are non-metallic dielectric coated to provide very high reflectance. Even the output mirror is around 99% reflective.

HeNe lasers may have linear or random output polarisation. Inherently, the output is random, but a polarised output can be produced by terminating the discharge tube with one or two Brewster windows, which are glass plates mounted at an angle such that their transmittance for one polarisation is 100% and about 96% for the other. This large loss introduced for the one polarisation results in nearly all of the power going into the other polarisation with no net loss in output power.

HeNe lasers are used in holography, scanning systems, alignment systems, measurement devices, vision systems and some line-of-sight communications applications, to name just a few of the dozens of successful applications.

Diode lasers

The diode laser is a solid laser – actually a single crystal semiconductor. Commercial devices are compound semiconductor alloys of the III-V type, meaning the main constituents came from the third and fifth column of the periodic table, e.g. gallium-arsenide (GaAs) and indium-phosphide (InP).

Fig. 3.3 is a sketch of a cross-section of a stripe-geometry double-heterostructure diode laser.

GaAs/AlGaAs lasers emit wavelengths in the 0.78–0.905μm range depending on composition. InP/InGaAsP type lasers emit in the 1.1–1.6μm range depending on composition. Efforts are underway to develop commercially feasible visible-emitting diode lasers.

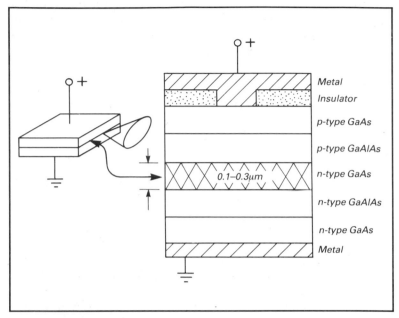

Fig. 3.3   Double heterostructure diode laser

There are many designs of diode lasers, but Fig. 3.3 illustrates the basic concepts. The lasers are manufactured (crystal growth) by liquid-phase epitaxial growth. The basis of operation is the injection of electrons and holes into the active region (*n*-type GaAs) by a large forward current, where electrons and holes recombine and emit photons. Parallel faces of the chip act as a resonator. The purpose of the ternary layers (AlGaAs) is to contain light as well as electrons and holes in the active region by means of reflection. The additional GaAs layers provide superior heat conduction. The purpose of the metal stripe geometry is to confine the current to a narrow region to improve heat dissipation and to make the divergence of the output beam more symmetrical. The narrowness of the active region from which the light is emitted results in high diffraction spreading (up to 40°). Outputs are still oval with divergence angles of 10° × 40° possible. This can be corrected through the use of cylindrical lenses.

Power output from diode lasers varies from 1mW for single lasers to 0.5W for phase-locked arrays of diodes built in a common substrate. Efficiency may be as high as 20% for diode lasers. Diode lasers can be operated CW or pulsed at high repetition rates.

Diode lasers are used in fibre-optic communications, printing, recording, optical discs, scanning, and even as a means of pumping (exciting) other lasers.

Nd-YAG lasers

The Nd-YAG laser (Fig. 3.4) is a solid laser in which Nd is the lasant and YAG the host that emits at a wavelength of 1.06μm (near infrared). YAG is an oxide with the chemical formula $Y_3Al_5O_{12}$. Power levels range from less than 10W to over 600W average power output. Up to 400W, these lasers can be operated CW, but most are pulsed for high peak power. For marking and engraving applications, CW Nd-YAG lasers are Q-switched by acousto-optic modulators to produce peak powers of over 1000 times the average power at rates of up to 10,000pps. The overall efficiency of Nd-YAG lasers is around 2%.

Nd-YAG lasers are pumped by Xe or Kr lamps; Xe is best for pulsed operation, Kr for CW operation. Optical materials are fairly conventional except for resonator optics

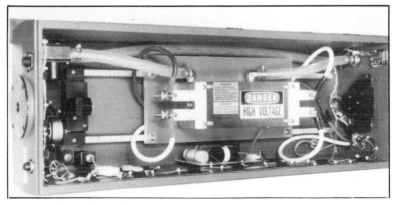

Fig. 3.4 Nd-YAG laser (Courtesy: Raytheon Corp., Laser Products Division)

which must be dielectric mirrors. Focusing lenses are generally BK-7 glass – a borosilicate crown glass known for its stability and high transmittance from 0.4 to 1.4µm.

Nd-YAG lasers are used for spot welding, seam welding, hole piercing, cutting and even magnetic domain refinement in transformer core sheet steel.

Nd-glass lasers

Nd-glass lasers, like the Nd-YAG laser, emit at 1.06µm, but with a much greater bandwidth because of the interaction of the Nd ions and the glass host material. Due to the poor thermal characteristics of the glass, these lasers are pulse operated with energy per pulse up to 60J and repetition rates of about 1pps. The overall efficiency of Nd-glass lasers is about 5%.

Nd-glass lasers are used in manufacturing primarily for hole drilling and spot welding. Glass laser rods can be either circular or rectangular in cross-section. With a rectangular rod, with a width-to-thickness (aspect) ratio of about 3:1, convenient stitch welds can be made since the output beam is rectangular and has essentially the same aspect ratio.

Fig. 3.5 shows a two-headed Nd-glass laser system (only one head is shown). Because of the low repetition rate of Nd-glass lasers, two laser heads can be alternately fired from the same power supply.

CO$_2$ lasers

The CO$_2$ laser undoubtedly deserves the title 'workhorse' of the laser materials processing industry. In spite of the inherent problem of high metallic reflectance at 10.6µm, the CO$_2$ laser wavelength, this laser has the widest diversity of applications and materials to which it can be applied. The number of distinct designs and range of power levels commercially available are also much greater for the CO$_2$ laser than other processing lasers.

CO$_2$ lasers vary in power output from a few watts to 15kW and higher. Output modes may be nearly pure Gaussian

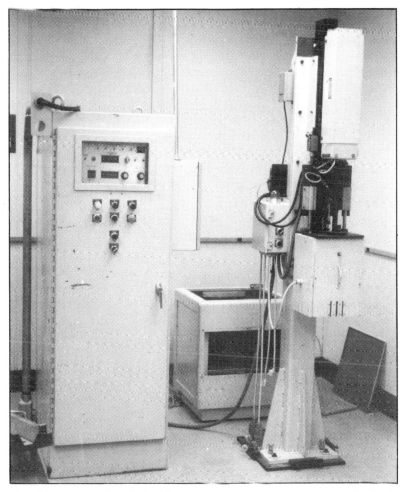

Fig. 3.5 A two-headed Nd-glass laser (only one head is shown). From left to right: power supply and controls, heat exchanger, laser stand with head and enclosed work area

(TEM$_{00}$), doughnut (TEM$_{01}$*), higher order (e.g. TEM$_{57}$) or of the unstable variety, particularly for multikilowatt lasers. Excitation may be dc, ac or ac with a dc bias. The highest ac frequency is about 10kHz.

CO_2 lasers generally operate with a mixture of CO_2, N_2 and He. CO_2 is the lasant, but N_2 is essential in the excitation process and is similar to He in a HeNe laser. The main difference is that the CO_2 laser is a molecular laser, meaning that molecular vibrations produce the light rather than electronic transitions. N_2 molecules, excited to vibrate by collisions with electrons in the discharge, transfer their energy to CO_2 molecules which then undergo transitions to a lower energy vibrational state giving off photons. He acts as an internal heat sink by helping to get CO_2 molecules back into the ground state so they can repeatedly interact with N_2 molecules to keep the lasing process going.

The efficiency of CO_2 lasers, the best of the processing lasers, is around 10%.

The most common design types are now described in a general way and their output characteristics and approximate power ranges presented.

Slow-axial-flow

This is the earliest commercial design and is still available from many companies. Fig. 3.6 illustrates this type of CO_2 laser.

An axial discharge is applied over a narrow bore tube up to a distance of about 1.5m and gas is flowed slowly at around 20–30Torr along the tube. The gas may be recirculated or expelled. The laser tube is surrounded by a cooling jacket filled with a dielectric liquid for lasers over 100W.

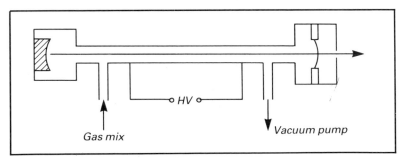

Fig. 3.6   Slow-axial flow design

Because cooling is by thermal conduction to the liquid, the tube bore must be small. This limits the amount of active material and thus the power output is limited to 50–70W per metre of tube length. Higher power levels are obtained (as can be seen in Fig. 3.6) by folding the beam back and forth in a variety of ways, depending on the manufacturer.

Although the narrow bore limits power, it has a useful side-effect in that it has a natural tendency towards low-order mode output. This type of laser commonly operates TEM_{00} or TEM_{01}^*.

The power range for this type of design is about 50–1200W. Such lasers can be operated CW or in rapid pulse fashion to achieve peak power three to four times higher than the CW power. Power output from a given laser is variable over a wide range with good stability.

Sealed-off CO_2

There are a number of sealed-off CO_2 lasers available up to 100W. They generally operate off 110V, but may require water cooling if operated continuously. The tubes are rechargeable after a few thousand hours of use.

One version of the sealed-off lasers is the waveguide laser (Fig. 3.7). In this design a ceramic tube with an extremely narrow bore is used so that the tube literally acts as a waveguide. They are frequently tunable over a small wavelength range, due to the many rotational levels of the CO_2 molecules.

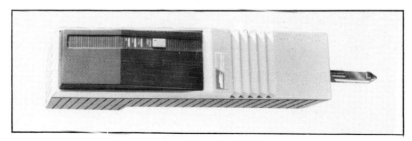

Fig. 3.7 A waveguide laser (Courtesy: Ebtec Corp.)

The output mode of sealed-off CO_2 lasers is nearly pure Gaussian and both CW and pulsed operation are available.

Sealed-off CO_2 lasers are compact and are very practical for light manufacturing operations such as cutting or drilling holes in polymeric or ceramic materials and soldering.

Fast-axial-flow

In fast-axial-flow CO_2 lasers (Fig. 3.8) a dc or pulsed electrical discharge is applied axially to a tube about 1m in length and the gas mixture is blown through the tube axially at about 60m/s or faster. The gas mixture is recirculated through a heat exchanger. Since this is a more effective method of removing heat from the gas mixture (referred to as convective cooling), the tube inner diameter can be much larger than in the conductively cooled slow-flow or sealed-off designs. Hence, more active medium is available for a given tube length. Power output of 600W per metre of tube length is common. This type of laser is available in power levels of 600–3000W. Near Gaussian or symmetrical low-order mode operation is attainable with modern versions of this design.

Fig. 3.8   600W fast-axial-flow laser and control console. (Courtesy: Lumonics Material Processing Corp.)

One of the major advantages of this design is its compactness coupled with the capabilities of wide power adjustment, pulsed or CW operation and good mode quality.

Transverse-fast-flow

There are several variations of the transverse-fast-flow CO_2 laser. In this type of laser (Fig. 3.9) the gas mixture is blown through the discharge region transverse to the beam. The beam may be folded back and forth through the discharge or, as in at least one version, the laser head has four separate legs arranged in a square fashion; each leg has its own discharge and gas flow. Gas flow rates vary from about 60m/s for the traditional designs to supersonic for some newer designs.

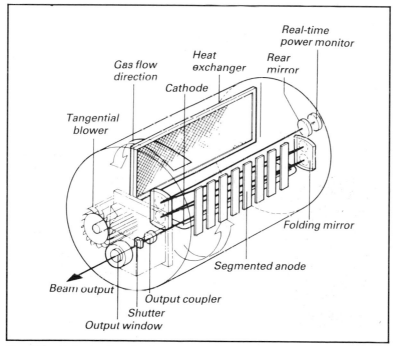

Fig. 3.9 Schematic of a transverse-fast-flow laser
(Courtesy: Spectra-Physics, Industrial Laser Division)

Excitation in transverse-fast-flow lasers may be dc or ac, including dc biased ac with frequencies up to 13.57kHz. The discharge can be either transverse to the gas flow or parallel to it.

For power levels up to 6kW, stable resonator designs are common, with unstable resonators optional in some cases. Above 6kW, unstable resonator operation is preferable because of the absence of a transmissive output mirror.

The large cavity configuration of transverse-flow lasers lends itself to higher order mode operation for stable resonators. However, low-order mode operation is possible with proper resonator design and stable discharge operation.

Transverse-fast-flow designs can be extremely compact with power outputs in excess of 3000W per metre of laser head length being common. Some particularly compact designs have resulted from the introduction of ac discharge operation.

Lasers of the transverse-fast-flow type range in power output from 800W to over 15kW in commercial versions.

Transverse-excited-atmospheric (TEA) lasers

This is a CO_2 laser design in which the sealed-off gas mixture is maintained at a high pressure (near atmospheric) and electronic pulses are applied to it through a system of pin electrodes transverse to the discharge. Such lasers are capable of pulse energy up to around 150J with pulse lengths varying from the nanosecond range to 100ms with pulse repetition rates of about 0.1–1000Hz.

These lasers are used primarily in marking and other applications where a large area beam of high pulse energy is required.

Excimer lasers

The term excimer stands for excited dimer, where a dimer is a compound of two identical species. Excimer lasers are not actually based on excited dimers, but excited complexes (exciplex) such as argon fluoride (ArF), krypton fluoride

(KrF), xenon chloride (XeCl) and xenon fluoride (XeF). These molecules can exist only when the noble gas atom is in an excited state. The bond between the excited noble gas atom and halogen atom is very strong, but the lifetime of the excited state is short, being of the order of nanoseconds.

The molecular species is formed in a pulsed electrical discharge in a resonator much like any other laser, except that the cavity is generally rectangular, as is the output beam, and the discharge is transverse. Gas pressure is near atmospheric. In fact many of the commercial excimer lasers can be used as transverse-excited-atmospheric CO_2 lasers.

When the noble gas atom returns to the ground state, the molecule dissociates, releasing the binding energy as a photon. This process can be stimulated so amplification is possible. The gain is very high for excimer lasers so relatively little feedback is required. The transmittance of the output coupler (output fraction for unstable optics) is usually about 0.50.

The wavelengths of the various excimer lasers are: ArF, 0.193μm; KrF, 0.248μm; XeCl, 0.308μm; and XeF, 0.351μm. These range from the vacuum ultraviolet to the near ultraviolet. Average powers of over 100W and pulse repetition rates of up to 1000pps are available depending on the particular excimer gas used.

Many excimer lasers are designed to use all of the gases mentioned, including CO_2 as a TEA laser. However, care must be taken to properly condition the laser between gas changes (this is called passivation). Halogens are extremely reactive and form compounds with materials in the laser. The laser must be heated to break these compounds and an inert gas such as He or Ar is flushed through the laser to remove contaminants.

Because of the high energy of excimer laser photons, direct interaction with bonds in polymeric materials occurs. This has opened up a myriad of applications in surgery and polymer processing, especially in the microelectronics industry. Less thermal damage occurs and sharper, cleaner edges can be cut.

Chapter Four

Identifying laser applications

THE PURPOSE of this chapter is to present the advantages and disadvantages of lasers in manufacturing applications along with some of the things to look for when trying to identify potential applications. Circumstances that may ameliorate the effect of the disadvantages are also discussed.

It is not difficult to argue that the laser, in its many different forms, is the most versatile manufacturing tool in existence. It is used for a variety of surface modification applications such as heat treating, cladding and alloying. It has numerous material removal uses such as cutting, hole piercing and marking. The laser has found extensive use for joining operations such as welding, brazing and soldering. It would be a hopeless task to compare the laser with all of the possible alternative techniques for doing the applications mentioned. Some relevant comparisons will be made to highlight advantages or disadvantages of the laser in certain applications.

Advantages *Read this!*

In this section a number of advantages of the laser are discussed in general terms. Some mention of specific applications may be made to clarify the discussion. In the chapters on applications, the significance of these advantages will be given more detailed attention. The advantages of the laser are direct or indirect results of the unique properties of laser light. All such properties are not important in all applications. In fact, in a few cases a unique property may be a disadvantage.

The major advantages are:

- High monochromaticity.
- High coherence.
- Small beam divergence (high radiance).
- Can be focused to small spot.
- Easy to direct beam over considerable distances.
- Small heat-affected zone (HAZ) in material processing.
- Propagate through most gases.
- Can be transmitted through transparent materials.
- No inertia or force exerted by beam.
- Easily adapted to computer control or automated manu- facturing systems.
- Not affected by electromagnetic fields.
- Wide range of power levels, mW to tens of kW.
- Wide range of pulse energies, µJ to tens of J.
- Wide range of pulse repetition rates, pulse lengths and pulse shapes.

Not all of the advantages directly associated with beam properties are true for all lasers. For example, diode lasers have divergence angles of 10–40°, excimer lasers have very poor spatial coherence, and Nd-glass lasers are not very monochromatic (large bandwidth). This does not negate the usefulness of these lasers in a variety of applications.

High monochromaticity is not important in materials processing applications, except potentially in the polymer and semiconductor industries. The same goes for coherence. The lack of coherence of excimer lasers is actually an advantage for photolithographic work in the semiconductor industry, but that is not, strictly speaking, a processing application. There are few transparent materials for trans- mission of CO_2 or excimer laser radiation.

High spatial coherence presents a problem called speckle, which gives laser light a characteristic grainy appearance

Mascot → 吉祥物

Chronology 按照时间顺序

Judging 判断

Some enterprises may find these as a good investment with low budget. As a result.

when viewed by reflection from most surfaces. On the other hand, speckle interferometry, discussed in a later chapter, is a useful industrial laboratory tool for determining small distortions of parts.

In low-power applications, such as holography and measurments, good coherence and monochromaticity are major factors along with high radiance and the potential for very good frequency stabilisation, which is particularly important for Doppler velocimetry and interferometric displacement measurement.

In high-power applications, high radiance is particularly important because that is what provides the capability of focusing the beam to a small spot. Several of the other advantages follow from the high radiance, e.g. propagation over large distances, large energy or power densities and small HAZ.

Disadvantages

Lasers are not all things to all applications. To be cost-effective an application must take advantage of the unique characteristics of the laser. When these types of applications are identified, the disadvantages become incidental. Nevertheless, they must be considered in the decision-making process when determining the viability of using a laser for a specific application.

The major disadvantages are:

- High capital cost.
- Low efficiency.
- High technology.
- Safety.
- Operator training.

These disadvantages are not pertinent to all applications. Low-power HeNe lasers can be purchased for a few hundred dollars, whereas a multikilowatt CO_2 laser may cost several hundred thousand dollars. Frequently the price of the laser is a small fraction of an overall system cost. If the use of a laser increases productivity through improved quality, higher throughput of elimination of process steps, the high capital cost may be justifiable.

Efficiencies vary from 0.1% for HeNe lasers to 10–15% for CO_2 and 10–20% for diode lasers. Efficiency, however, is not important if there is no other way to do the job or if the total power consumption is low.

Lasers are normally perceived as high tech devices. However, most of the technology in lasers is fairly standard, particularly the electronics, controls and associated computer equipment. The resonator and optics (internal and external) are generally the only unfamiliar parts of a laser or system that incorporates a laser.

Safety is an important factor in the use of lasers. Not only can the beam be a direct hazard, but frequently the fumes or other debris produced during material removal are hazardous. Fire and electrical shock are also potential hazards. There are many good publications and numerous courses available on the subject of laser safety. For a very nominal cost one or more people can be trained in this area. One of these people should be designated as the Laser Safety Officer (LSO) and be given the responsibility and authority for ensuring the safe use of lasers.

The training required for laser operators is minimal. Usually, more skill is required to operate associated equipment such as CNC or material-handling equipment. However, someone with supervisory responsibility, such as the engineer in charge, should have an educational background and experience (if possible) in the use of the types of laser under that person's supervision. Simple but fundamental mistakes are frequently made because of a lack of understanding of the laser and/or its associated optical system.

Identifying laser applications

In this section some very general remarks will be made concerning what to look for in determining whether or not a laser should be used in a particular application.

Low-power applications

In this subsection we are talking about lasers generally in the 1–50mW range, usually HeNe or diode lasers. Applications of this type are easy to decide on because invariably there is *no other way* to do the job. The characteristics of lasers which make them useful in this type of application are high monochromaticity (frequency stability), coherence and high radiance (low beam divergence).

Interferometric applications such as holography, speckle, measurements, and Doppler velocimetry require good spatial and temporal coherence, as well as varying levels of frequency stability and good mode quality. Laser interferometry is used to make measurements or to control machines and motion systems to within a fraction of a micrometre. The small beam divergence of lasers gives them unique capabilities in alignment applications whether for machine set-up or building construction. The high radiance permits the use of lasers for accurate triangulation measurements of absolute distance for both measurements and control of machines such as robots. Also, due to the higher radiance, laser light can be structured, such as formed into a line of light, for vision applications to part location, orientation and/or identification.

These types of applications would not be feasible in an industrial setting using conventional light sources. In most cases, no other comparable technique exists.

High-power applications

Materials processing applications are more difficult to analyse than most low-power applications. Most processing applications done by the laser could be done by some other technique. There are obviously many competing technologies in welding, heat treating and material removal. What one has to look for is some unique aspect of the application which would make good use of one or more of the laser's unique capabilities.

In welding applications the major factors to look for are:

• Minimal distortion due to heat input.

- Cosmetically good weld bead.
- One-sided access.
- Narrow weld bead with deep penetration.
- No special joint preparation or filler metal required.
- Many dissimilar metals easy to weld.

There are limitations to laser welding, as with any process. Spot welding with pulsed solid-state lasers (also seam welding) is limited to about 1–2mm penetration, and continuous seam welding with multikilowatt CO_2 lasers is limited to around 1cm penetration. High carbon steels and certain other alloys are difficult to weld without cracking because of the rapid heating and cooling involved. Reasonably good fitup between parts is required for successful laser welding.

In cutting applications the following are the major considerations:

- Easy to cut complicated two- or three-dimensional contours.
- No tool contact.
- Small heat-affected zone (HAZ).
- Small runs with variable shapes.
- Smooth as cut edges.
- Readily adaptable to robotic or CNC systems.
- Easily integrated into flexible manufacturing systems.
- Cuts with ease refractory metals and many other materials that are difficult to machine.

The major drawback to laser cutting is thickness of the material to be cut. Nd-YAG lasers can cut up to about 5mm thickness and CO_2 lasers can cut up to about 20mm at reasonable speeds. Large amounts of assist gas are required. If oxygen is used, the operating cost can be quite high. If air can be used, this cost is virtually eliminated.

Most of the statements concerning cutting apply to hole piercing (drilling) with solid-state lasers. Added to the list should be the fact that holes can be drilled at large angles to the surface. Hole depth-to-diameter ratios (aspect ratios) of 20:1 are common. High repeatability and accurate placement are achievable.

The major drawback to laser hole piercing is that the holes are tapered and some recast material is usually present in the entrance side, and dross due to oxidation may stick to the exit hole.

Applications in microelectronics may require specific wavelength characteristics or the small focused spot size achievable with lasers.

Marking applications utilise the high radiance and flexibility associated with the laser beam for either beam manipulation using computer-controlled galvanometer mirrors or a masking technique.

General characteristics

There are a number of characteristics, some of which have already been mentioned, which apply in a general way to all applications.

Flexibility is a predominant one. The ease with which laser beams can be manipulated and transported make it possible to process rapidly extremely complex shapes in two or three-dimensions and to get into areas of parts where other tools could not be used. Short to moderate length runs of a few thousand parts where different designs or different parts are processed are frequently suitable applications.

Whether used in surface modification, joining processes or material removal, the heat input to the part is generally low and thermal distortion or chemical changes are minimal. The extremely high radiance of lasers is pertinent to virtually all manufacturing applications of lasers.

Chapter Five

Laser surface modification

LASER surface modification (LSM) refers to any process in which the heat produced by the interaction of the laser beam and the surface of a material is used to bring about a beneficial alteration in material properties. The purpose may be to increase wear resistance, corrosion resistance or strength.

The techniques described in this chapter are laser-chemical-vapour deposition (LCVD), cladding, alloying, surface hardening and surface melting. The processes are described and the desired result is discussed. Some of the specific advantages of using the laser in these applications are also pointed out.

The set-up used for LSM is relatively simple. The beam is directed perpendicular or nearly perpendicular to the surface to be treated, and either the part or the beam is translated at the appropriate speed to give the desired area-coverage rate. The spot size used in most LSM work is fairly large. A simple way to achieve this is to defocus by a certain amount. Unfortunately this alone does not always produce a sufficiently uniform beam. A higher order mode is generally better than a TEM_{00}, but these are sometimes hard to maintain and are not always sufficiently symmetrical.

Successful methods for overcoming the drawbacks associated with laser beams are segmented mirrors (Fig. 5.1) and

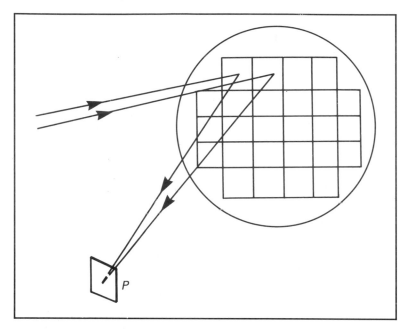

Fig. 5.1   Segmented mirror

beam scanning. Each segment of the mirror produces an image of the portion of the beam striking it in the image plane at P. When about 30 such images are superimposed at P, the result is a remarkably uniform power distribution regardless of the condition of the beam illuminating the mirror.

In scanning techniques the beam is oscillated or rotated to produce an effectively uniform power distribution. Patterns such as those shown in Fig. 5.2 are produced. In dithering, the beam is rocked back and forth as the part moves, to produce a 'sawtooth' pattern. If galvanometer mirrors are used, a rectangular scan pattern can be produced by rapid scanning. Part movement provides the necessary coverage. Rotation of the beam is accomplished in a manner similar to a process called 'trepanning', used to cut holes. The laser spot rotates as the part moves, producing a spiral path on the part. With rapid spinning the treated path is very uniform.

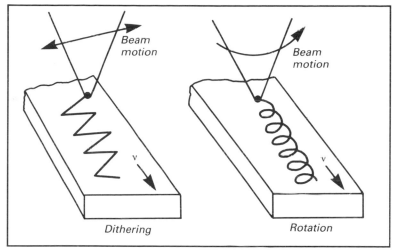

Fig. 5.2 Beam scanning (v is velocity)

Laser-chemical-vapour deposition

Laser-chemical-vapour deposition (LCVD) is a process whereby the laser is used to heat a substrate so that a vaporous material can be deposited on it. In some cases the laser is also used to vaporise the material to be deposited.

LCVD is used for many purposes, including improved wear resistance and corrosion resistance. In the semiconductor industry it is used extensively for deposition of a wide variety of materials such as dopants, SiO_2 and polycrystalline silicon (polysilicon).

The major advantage of using the laser for heating the substrate is for the selective deposition of the material. If large areas are to be deposited, the laser is an inappropriate heat source. If the deposition needs to be highly localised, either because of material cost saving or because of the complex small geometries required in integrated circuits, the laser may very well be a viable cost-effective heat source.

Cladding and alloying

Cladding and alloying are similar processes in that in both cases an appropriate powder is laid down on the surface and is melted by the heat from the laser beam. In cladding,

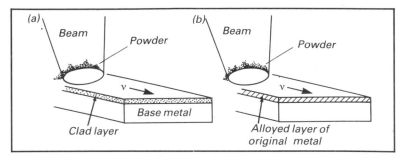

Fig. 5.3   (a) Cladding and (b) alloying

sufficient heat is put into the part to melt the powder and to produce a metallurgical bond between the solidified material that is added and the base metal surface. This process is also known as hardfacing.

In alloying, the base metal is melted to a substantial depth and the alloying material is mixed into the base metal to produce a new alloy. These two processes are schematically depicted in Fig. 5.3.

Cladding or alloying may be used for a variety of purposes such as wear resistance, corrosion resistance and improved impact strength. Usually the most appropriate applications are when only selected areas, such as valve seats, require treatment. Nevertheless, some fairly large area cladding applications on oil-drilling tools have been reported because of the high uniformity of the clad layer that can be achieved with minimal dilution by the base metal (Fig. 5.4).

Fig. 5.4   Laser-clad plate (Courtesy: Spectra-Physics, Industrial Laser Division)

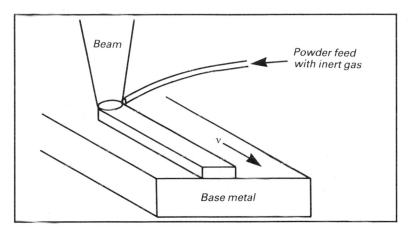

Fig. 5.5 Powder feed method for cladding and alloying

The powder may be applied prior to irradiation by the laser, frequently in some sort of organic binder that burns off during the heating proccss. The binder holds the powder together and sticks it to the substrate. In many cases the powder is blown onto the part at the point of incidence of the beam, as depicted in Fig. 5.5.

Surface hardening

The laser, particularly the CO_2 laser, provides a unique tool for surface or case hardening when selective hardening and/or minimal distortion is important. The most common reason for laser hardening is to improve wear resistance, but impact strength and fatigue strength can also be improved as a result of residual compressive stresses that are introduced into the surface. Base metal toughness can be retained while producing a hardened layer on the surface.

Only materials which are hardenable by conventional heat treating can be hardened with the laser. Cast iron and steel with greater than 0.2% carbon are hardenable. Pearlitic cast irons are easier to harden than non-pearlitic types, because of the fine distribution of the carbon. However, the wear resistance of non-pearlitic cast irons can be improved substantially by laser surface treatment.

In laser surface hardening, a defocused or otherwise smoothed-out beam is impinged on the metal surface and either the beam or the part is translated. A power density of a few thousand watts per square centimetre usually suffices if the intent is to heat the surface without melting. Metallic surfaces exhibit quite high reflectance at $10.6\mu m$, so coatings are generally used to help couple the beam power into the metal surface. Such coatings as Lubrite (manganese phosphate) and other phosphates work quite well and may already be on the part for another purpose. Graphite sprays and flat black paint also work well. The most suitable coating depends on the situation and the power level of the laser. Thicker coatings are required at high power levels so that the coating does not burn off before the beam has left a given spot on the part. The heat treatment process is illustrated in Fig. 5.6.

The heat produced by the incident laser beam raises the surface temperature of the metal to some point generally below the melting point of iron, around 1500°C depending on the alloy. It should be realised that the laser light does not penetrate the metal to a significant extent (less than $1\mu m$), so the heating takes place strictly at the surface. Thermal conduction carries heat into the metal, causing the temperature to rise. Each point in the beam path reaches some maximum temperature which decreases with depth

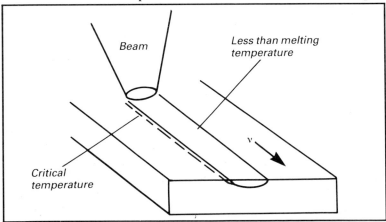

Fig. 5.6   Heat treating

Fig. 5.7   Laser-hardened track on rod of H-13 tool steel

into the metal. At some depth below the surface, usually
0.1–1.5mm, the maximum temperature reached will be the
critical temperature for transofrmation hardening – the
so-called austenitic temperature. This temperature is about
730°C for carbon steel. If the metal part is large compared
with the beam diameter, thermal conduction will result in
sufficiently rapid cooling to provide self-quenching. The
cooling rate in laser surface hardening can easily be
thousands of degrees per second.

Fig. 5.7 is a photograph of a laser-hardened track on a
steel shaft. The laser-hardened track is shinier than the base
metal because its microstructure (mostly martensite) is not
as readily attacked by the etchant as the softer base metal.
Fig. 5.8 contains micrographs of heat treat tracks in 1040
steel, and Fig. 5.9 is a plot of hardness (Rockwell-C scale)
versus depth of the centreline of a laser-hardened track.

In laser surface hardening of non-pearlitic cast iron it is
not possible to achieve a uniformly hardened track, because
the carbon is primarily in the form of graphite nodules or
flakes. Transformation hardening requires the diffusion of
carbon into nearly ferrite (nearly pure iron). If the carbon is
finely dispersed, as in pearlitic iron or steel, sufficient
diffusion occurs rapidly enough for uniform hardening to
take place. In a nodular cast iron, hardening will occur in a

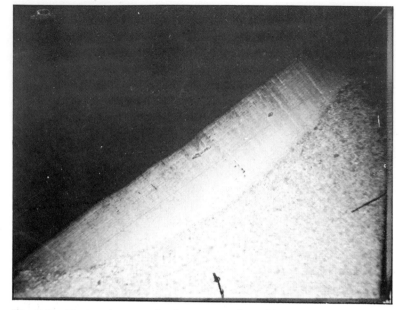

Fig. 5.8   Photomicrograph of cross-section of laser-hardened track in 1040 steel

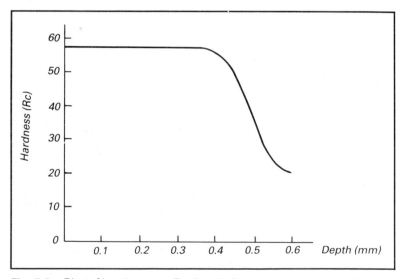

Fig. 5.9   Plot of hardness on Rockwell-C scale versus depth of laser-hardened track

small region around the nodules (for as far as the carbon can diffuse in the short time the temperature is elevated). This region around the nodules is extremely hard and provides a hard bearing surface, somewhat like impregnating a surface with hard particles of a material such as silicon-nitride. Essentially, the result is a soft matrix of iron with islands of hard martensite and iron carbide (cementite) surrounding the graphite particles.

Surface melting

There are two regimes of laser surface melting which might be referred to as slow melting and rapid melting or solidification.

In the first case, a highly defocused or integrated beam is scanned across the part at a speed which results in substantial melting of the surface. Using conditions similar to a laser heat treatment set-up, it is possible to achieve melt depths of over 0.5mm using a 2.5kW CO_2 laser, a track width of about 6mm and a speed of 2.5cm/s. The heat-affected zone beneath the melt layer is about 1.5mm thick at its deepest point. The purpose of this melting is to homogenise the microstructure. Fig. 5.10 shows micrographs of conventionally hardened and laser surface-melted H-13 tool steel.

Most of the nitrides and carbides were dissolved during the melting process. The solidification rate is less than

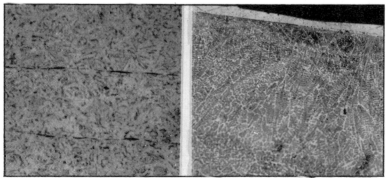

Fig. 5.10   Conventionally hardened and laser-melted H-13 tool steel

1000°C/s, but this is still rapid enough to prevent precipitation of these phases. Hence, a unique non-equilibrium homogeneous microstructure is produced.

The presumed advantage of this process is improved resistance to thermal stress cracking, which is believed to initiate at the hard carbide and nitride phases. The disadvantage of this process is that tensile stresses are produced in the melt layer. It may be possible, however, to remove those stresses by means of a subsequent heat treatment.

In rapid melting, a tightly focused beam is rapidly scanned across the material, producing a very thin melt layer only a few hundredths of a multimctre thick. Cooling rates of 1,000,000°C/s can be achieved. At such high cooling rates it is possible in some alloys to produce a non-crystalline (amorphous) or glassy layer. This process is referred to as laser glazing. The benefit of such a glassy metal is that it is generally more resistant to corrosion and may have superior wear characteristics. The main drawback to this process is that it takes a long time to cover a significant area.

Miscellaneous LSM

There are many LSM techniques that have been proposed or have found their way into production. Armco Steel uses Nd-YAG lasers to refine the magnetic domain structure of transformer steel. Other suggestions for LSM include glazing of ceramics and stress relief of ceramics by controlled introduction of surface cracks by laser heating. Some brake manufacturers are reported to be using lasers to irradiate metal-impregnated disc brake pads to prevent glazing of the pads during use, which produces a disconcerting squealing sound that makes the brakes appear defective.

Chapter Six

Laser joining

THE EXPRESSION 'laser joining' is used here to include welding, soldering and brazing. The major applications of laser joining, welding and soldering will be described in this chapter.

In soldering, the laser simply acts as a heat source to reflow the solder which has previously been placed on the part or parts. Two regimes of operation occur in welding, depending on the incident power and power density. These are referred to as thermal conduction welding and keyhole welding. These phenomena will be discussed along with the lasers that are used in joining processes and the advantages specific to laser joining.

Lasers used in joining

All of the primary industrial processing lasers are used for welding. Low-power Nd-YAG and CO_2 lasers are used for soldering.

For deep penetration ($\geqslant$2mm) seam welding at high speed the likely choice is the CO_2 laser. The power level that is needed can be estimated by the methods presented in the 'calculations' section of this chapter. (Experimental data in welding power, speed and penetration depths for most materials can be found in the references given at the end of the book for this chapter.) Calculations should always be verified by having a job shop or vendor do a feasibility study on the material and thickness that is required to weld.

Shallow-penetration seam welding ($\simeq$1mm or less), such as in electronic packages and relay cans, is usually done with a high-power pulsed Nd-YAG. These lasers run from 150 to 600W average power and are capable of producing energies of 50–60J per pulse. These lasers are also the best choice for high repetition rate spot welding.

For low-duty cycle spot welding ($\simeq$1pps), Nd-glass lasers should be considered because of their simplicity and low cost. Either round or rectangular rods can be used, depending on the spot shape desired. These lasers produce up to 60J per pulse and two heads can be operated from the same power supply for higher throughput.

Thermal conduction welding

For power levels under 1kW and power densities below 10^5W/cm^2, laser welding is predominantly controlled by thermal conduction. This means that the power is absorbed at the surface of the metal and diffuses into the metal. Melting occurs for a depth determined by the thermal characteristics of the material and the welding parameters. Obviously, there must be sufficient heat to raise the material's temperature above the melting point. Thermal conduction is not directional, so low depth-to-width (aspect) ratio welds are obtained. Fig. 6.1 is a photomicrograph of a typical thermal conduction laser weld.

Thermal conduction welding is satisfactory for many applications, including hermetic sealing of electronic packages. However, the depth of penetration that can be achieved in thermal conduction welding is limited to about 1–2mm by the thermal properties of metals. Attempts to achieve deeper welds either result in severe surface damage due to vaporisation or excessive melting of the surface.

All of the primary industrial processing lasers can be used in the thermal conduction regime. Seam welding with CW lasers of any type is chiefly a thermal conduction process for powers up to 500W. The exact point at which keyholing

Fig. 6.1 Photomicrograph of thermal conduction weld showing some keyhole effect

begins to play a significant role is debatable, but it does not appear to be a serious contributor up to about 1000W.

Solid-state pulsed lasers generally operate in the thermal conduction regime. However, even though average powers rarely exceed a few hundred watts, keyholing can be achieved for sufficiently high peak power-short duration pulses. The maximum penetration is still not much over 2mm. As the average power of solid-state lasers is increased, their ability to do deep penetration or keyhole welding will improve markedly.

Keyhole welding

The phenomenon which enables lasers to achieve deep penetration welds with high aspect ratios is referred to as 'keyholing'. This is a well-known phenomenon observed in electron beam welding before the laser was invented. In this phenomenon the vapour pressure of the heated metal overcomes the surface tension of the molten pool and opens up a cavity into the metal which allows the beam to 'see' into the bulk of the metal. Fig. 6.2 is an illustration of this phenomenon.

Since the beam is able to penetrate the metal to some depth relatively unimpeded, less heat is lost laterally by thermal conduction. In the 1–3kW range, typical welds are the results of a combination of thermal conduction and keyholing. Above 5 or 6kW, the welding process is almost entirely keyholing. Fig. 6.3 is a photomicrograph of a weld made with 2.5kW from a CO_2 laser. The widening of the upper part of this 'nailhead'-shaped weld nugget is due to thermal conduction. The narrow lower part is almost entirely the result of keyholing.

In thermal conduction welding a great deal of energy is lost due to reflection. Although the reflectance of metals decreases as their temperature increases, molten metal is still highly reflective, especially for 10.6μm radiation. The keyhole acts as a light trap, becoming virtually a blackbody

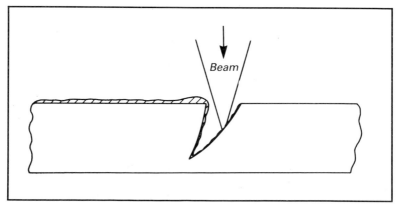

Fig. 6.2   Keyhole phenomenon in cross-section

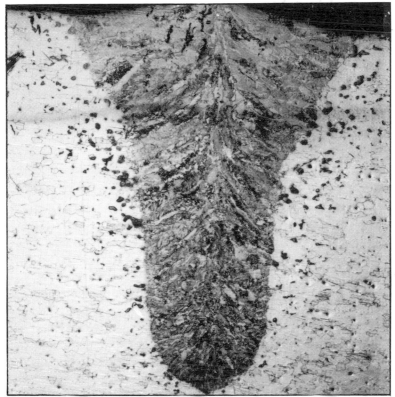

Fig. 6.3   Photomicrograph of weld nugget showing the effect of keyholing

at multikilowatt power levels for CO_2 lasers. Thus, keyholing improves depth of penetration and weld efficiency. Weld efficiency is defined as the ratio of energy used to actually produce the *desired* nugget divided by the total energy delivered to the part.

In the keyhole welding process, molten metal flows around the keyhole and fills the hole back up after the beam has passed over. In continuous seam welding, if the beam is turned on, or off, rapidly, such as by shuttering, a dimple will be left at the end of the track providing clear evidence of the presence of the keyhole. This dimpling can be eliminated by ramping the power up at the beginning and end of the weld cycle. The lower power level is called the simmer power. A simmer power of 1kW or less and a ramp time of 0.1–0.2s will generally eliminate the dimples.

Because the keyhole provides an opening for the laser beam to penetrate the metal, the optimum focal position is usually below the surface. Both weld nugget shape and depth depend on focal point position. The optimum focal point position varies with materials and joint configuration, but frequently lies in the range 0.5–1.5mm below the surface.

Laser weld joint configurations

The basic limitations on joint design are thickness and beam access. The laser beam must have a 'line of sight' to the joint. The thickness of material that can be welded depends on the laser power level and thermal characteristics of the material. Thermal conduction welding is limited to thicknesses of 1–2mm, whereas penetration depths of 1.0cm or more can be achieved with commercial multikilowatt CO_2 lasers.

Butt welds and lap welds are standard laser weld configurations, though attention must be paid to fit-up conditions. Other joint configurations suitable for laser welding are T-joints (fillet welds) and coach joints (Fig. 6.4)

In T-joints or fillet welding (these are essentially the same) the beam is brought in at an angle of about 20°

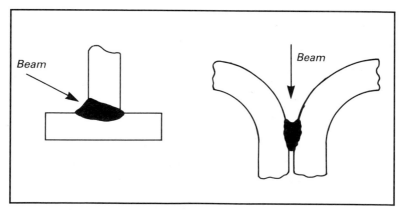

Fig. 6.4   Weld joint configurations

relative to the seam. If this is done correctly, the beam will be guided, to a certain extent, by the seam, so that complete fusion of the two parts occurs. The bead opposite the beam entrance side will be on the seam. If the beam angle is too steep, the bead will go below the seam, leaving a portion of the seam unwelded, which will adversely affect joint strength.

In the coach joint, multiple reflections off the sides of the joint guide the light down into the seam until sufficient absorption occurs to cause the metal to melt and fuse the two pieces together. Coach joint welding will be sensitive to beam size, depth of focus and curvature of the pieces being joined.

When welding pieces together which have different thermal behaviour, such as butt welding thin materials to thick materials or different alloys, the beam spot can be preferentially located more on one material than the other. Thus, more power can be put into the material that is more difficult to melt to improve fusion of the pieces.

Preferential spot location can also be used for the purpose of controlling mechanical characteristics of the weld bead. For example, when welding low-carbon steel to high-carbon, you can melt more low-carbon than high-carbon steel to dilute the carbon in the weld bead. If this is not done, a brittle weld may be produced.

Welding calculations

Lasers are used for both spot welding and continuous seam welding. In some cases seam welding is done by overlapping spot welds. Most spot welding is done with solid lasers with pulse lengths in the range of 1–10ms. a rough correlation of power, spot size, weld depth and pulse length can be made based on simple energy balance; in this approximation, heat loss by conduction, radiation and convection is ignored. If the quantity $\sqrt{\kappa t}$, where κ is thermal diffusivity and t is time, is small compared with the beam spot dimensions on the part, this approximation is fairly accurate. For example, for steel, $\kappa \approx 0.1 \text{cm}^2/\text{s}$, so for a 5ms pulse, $\sqrt{\kappa t} = 0.01$ cm, which is small compared to the spot size in most cases. The quantity $\sqrt{\kappa t}$ is a measure of the distance of heat diffusion in time t.

For the energy balance approximation, the energy required to produce a weld nugget of a given volume V is given by:

$$E = \rho V(C\triangle T + L_f) \tag{6.1}$$

where ρ is density, C is specific heat, $\triangle T$ is change in temperature and L_f is the latent heat of fusion. As an example, consider producing a 1mm deep nugget in steel. For a thermal conduction weld, the nugget will be approximately hemispherical so the volume is given by $\frac{1}{2}(4/3\pi r^3)$ where r is the nugget depth, or radius. For steel $C \simeq 0.44 J/g°C$, $L_f \simeq 247 J/g$ and $\triangle T = T_m - T_a = 1500°C$, where T_m is the melting temperature and T_a is the ambient temperature. Plugging all this into Eqn. (6.1) yields about 1.5J. This is the actual energy required to produce the weld nugget. It would be judicious to assume a reflectance of 0.5 and a welding efficiency of 0.5 which means that one-fourth of the energy incident on the part actually goes toward producing the weld nugget. Thus, the actual energy needed is 1.5J/0.25 = 6J. Some experimentation will be required to determine the optimum pulse length and pulse energy, but the calculated number provides a reasonable starting-point.

The energy balance approach can also be applied to continuous welding to obtain an approximate relationship between speed, power and depth of penetration. Eqn. (6.1) gives the energy balance relationship for a weld nugget of a given volume V. For continuous welding, we let $V = wld$, where w is width, l is length and d is depth of the weld bead. Then:

$$P = E/t = \rho wdV(C\triangle T + L_f)$$

where $v = 1/t$ is the weld speed and P is the power that actually produces weldment.

As an example, again consider steel in which we wish to produce a 2mm deep weld with an average width of 0.75mm. To do this at 4.25cm/s (about 100in./min) requires:

$$P = 7.86g/cm^3 \ (0.75mm) \ (2mm) \ (4.25cm/s)$$
$$(0.44J/g°C \ 1500°C + 247J/g) = 455W$$

Assuming that four times this much power is required to do the job, as in pulse welding, gives 1818W, which is not far from reality.

A dimensionless analysis which takes into account the thermal conduction characteristics of metals provides a better and more accurate understanding of the relationship between power, speed and weld depth:

$$d \simeq 0.15 P/T_m k(\kappa/VW)^{1/2} \tag{6.3}$$

where k is thermal conductivity, κ is thermal diffusivity and T_m is the melting temperature. Eqn. (6.3) gives remarkably good correlation for a wide variety of materials and welding parameters and applies to both conduction and keyhole welding conditions. As an example, consider welding steel with 2.5kW at 4.25cm/s with a weld bead width of 1mm:

$$d = 0.15 \ \frac{2.5kW}{1500°C \ (0.4W/cm°C)} \left(\frac{0.1cm^2/s}{4.25cm/s \ (0.1cm)} \right)^{1/2}$$
$$= 3mm$$

The predicted penetration is 3mm, which is definitely in the right range.

There are many factors which influence welding in addition to the material parameters. Beam mode, focused spot size and the type of cover gas and manner in which it is applied all have important effects on speed, penetration and shape of the weld nugget or bead. Even whether the laser is rapidly pulsed or operated CW makes a difference in seam welding. Improved penetration and weld quantity can sometimes be achieved with enhanced pulse welding. This is where a CO_2 laser is rapidly pulsed to achieve peak powers that are several times the average power, which may be 1–3kW.

Cover gases

Cover gases are used for a variety of reasons in laser welding. These include protection of focusing optics, control of weld plasma, and protection of weld bead from contamination and oxidation.

There are several ways of applying the cover gas. A flow of gas coaxial with the laser beam prevents smoke and mild spatter from depositing on the focusing optic(s). However, this type of cover gas arrangement will not prevent violent spatter from reaching the final focusing lens or mirror, even for relatively long focal lengths such as 25cm. To stop or minimise the damage to optics due to violent ejection of molten material, a strong crossflow of air is frequently employed. This technique must be applied carefully to avoid disruption and contamination of the weld bead by the airflow. The pressure drop caused by the Bernoulli effect can cause excessive removal of material from the weldment.

Off-axis cover gas can be uniformly supplied by a MIG welding nozzle such as that shown in Fig. 6.5. This type of nozzle projects a uniform stream of gas for several centimetres and when used with a heavy gas such as argon or nitrogen provides excellent coverage.

The most commonly used gases in laser welding are helium, nitrogen, argon and air. Carbon dioxide is sometimes employed, but yields poor penetration, especially with CO_2 lasers, where the penetration may be less than a third of that obtainable with the other gases.

Fig. 6.6 shows a laser-welding operation. The bright plasma is evident at the surface of the workpiece. The plasma is not generated directly by the laser energy, but it is supported by it – hence the term laser supported plasma. The plasma is caused by the heating of the workpiece which emits small particles, electrons and ionised atoms. Some of the laser energy is absorbed in this 'soup' causing additional heating and additional plasma to be generated.

Helium is generally accepted to be the best cover gas from the standpoint of minimising plasma absorption. This is because helium has a higher thermal conductivity and higher ionisation potential than the other gases. Unfortunately, the high cost of helium frequently makes its use in production prohibitive.

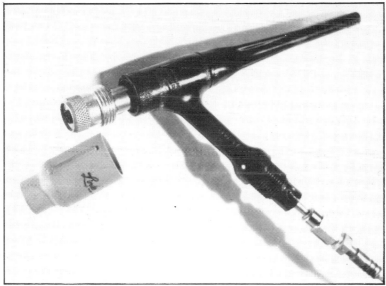

Fig. 6.5   MIG welding nozzle

Fig. 6.6   Laser welding

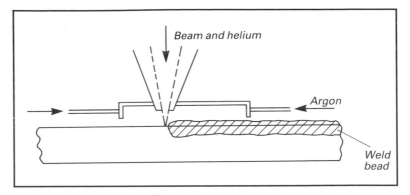

Fig. 6.7   Trailing cup shield gas arrangement

Argon has the advantage of having the highest molecular weight (except for CO_2) and therefore has excellent covering capability. In some applications where weld bead protection is critical, such as welding of highly oxidisable metals such as titanium or aluminium, helium is used coaxial with the beam and argon is applied by means of a 'trailing cup' as shown in Fig. 6.7.

Helium and argon are sometimes mixed to optimise depth of penetration and coverage. In non-critical welding, air is frequently used since it can be supplied by a compressor. Great care must be taken to ensure that the air is free of oil and moisture to prevent damage to the focusing optics. Nitrogen is relatively inexpensive and may be a reasonable compromise between helium and argon. However, if nitrogen is absorbed in the weld bead, embrittlement will occur.

A number of techniques for suppressing or controlling the plasma are in use. The simplest approach is to blow the plasma away from the weld spot with a carefully aimed stream of inert gas. Care must be taken not to disrupt the weld pool in this technique and precise aiming of the gas stream must be maintained for it to be effective. A technique which requires even more precise control is to aim a small stream of inert gas directly into the keyhole, thereby forcing the plasma into (or containing it in) the keyhole where the heat is transferred to the metal. The difficulties with this very effective technique are the precise alignment that must be maintained and damage from heat and spatter to the small tube used to direct the gas flow.

The ultimate plasma control is to weld in a vacuum. Under this condition, laser welding approaches the penetration depths achievable with electron beam welding at comparable power levels. Unfortunately, this is not very practical.

Advantages and disadvantages of laser welding

Specifically, the advantages of laser welding are:

- Flexibility.
- No filler metal, normally.
- Weld in most atmospheres.
- No vacuum needed.
- Weld many dissimilar metals.
- Deep penetration.
- Moderately high speed.
- No special joint preparation.
- Minimal part distortion.
- Small heat-affected zone (HAZ).

The laser is not magical and is not an answer for welding materials with poor weldability such as the 5000 series aluminium or aluminium to steel. However, it does do a superior job in many instances of welding dissimilar metals. A filler metal may be used when alloying is required to produce an acceptable weld.

Although special joint preparation is not usually done, edges need to be clean and straight for butt welding and in all cases attention must be paid to achieving proper fit-up. Lasers are not forgiving for poor fit-up conditions. For butt welds, if the gap between the parts is comparable to the focused spot size, most of the power may go through the crack. It is possible to defocus or use a long focal length focusing lens or mirror, but in any event if the gap is about 15% or more of the metal thickness, inadequate fusion between the two pieces may occur because the laser beam melts a relatively small amount of material.

In lap welding, a larger gap can be tolerated. However, this becomes a question of aesthetics and possible strength. For large gaps, severe undercutting occurs because of the small amount of material melted. Remember, the gap between the pieces has to be filled by metal melted in the top piece. Similar considerations apply to other types of joints.

The disadvantages of laser welding are:

- High capital cost.
- Moderate operating cost.
- Medium level technology.
- Requires special maintenance and operator skills.
- Special safety considerations.
- Limited penetration depth.

Obviously, there are many instances where the advantages and cost effectiveness of laser welding outweigh the disadvantages, particularly where minimal distortion and/or heat input are critical. Frequently, laser-welded parts can be used in the 'as is' condition because of the cosmetically superior weld bead and the low distortion that is produced.

Lasers have been in production for over 15 years, so one could say they are no longer 'high tech'. Safety precautions, though extremely important, are chiefly common sense. The training required for maintenance of lasers is *not* ominous, but it is important that technicians be trained in the proper handling techniques for laser components, especially optical components.

Laser soldering

In laser soldering, the laser simply supplies heat for reflowing solder which has already been placed on the part or parts to be connected. The solder may be in the form of preforms or solder ink that is printed onto wiring pads. Typically, power levels used for soldering are 10–20W, though CO_2 lasers with much higher power levels have been used.

Lasers are used for soldering because the heat input can be localised and minimal heating is required. Hot-plate or hydrogen flame techniques heat either the entire substrate or a large portion of it and this can damage sensitive electronic components. Also, if all components and wire bonding are not done simultaneously, solders of different alloy content must be used to alter their softening temperatures so that previously soldered components are not debonded.

With the precision that can be achieved with laser soldering, debonding is not a problem, even for very closely spaced leads, wires or components.

Whether Nd-YAG or CO_2 is the best choice for a given application depends on the absorption characteristics of the substrate and other packaging material that might be irradiated either directly or by scattered light. Ceramics, plastics and glasses are extremely absorptive at $10.6\mu m$. Metals, including solder, are not as absorptive of $10.6\mu m$ wavelength radiation (CO_2 laser) as of $1.06\mu m$ radiation (Nd-YAG laser). One situation in which the CO_2 laser is definitely superior is where a polymer coating must be stripped from the wires just before they are bonded to a terminal pad. A CO_2 laser can do both the stripping and the soldering.

Chapter Seven

Material removal

THIS chapter examines the processes of hole-piercing (drilling), cutting and marking with lasers. The types of lasers and systems used in these processes are described, and some of the advantages and disadvantages of various approaches are discussed, together with the effects of beam quality and polarisation.

Laser hole-piercing (drilling)

There are two methods of piercing holes with lasers: percussive drilling and the gas-assist process. In percussive drilling, the energy from the laser pulse causes material to vaporise rapidly enough so that molten, and sometimes solid, material is expelled from the hole due to the rapid high-pressure build-up.

In the gas-assist process, air or an inert gas is blown into the hole to aid the expulsion of molten material and solid particles. If the material is oxidisable, the oxygen in the air will react exothermically with the material, thereby enhancing the drilling capability of the laser.

Holes drilled with lasers typically have tapered sides and less-than-perfect roundness. Roundness can be maintained by using lasers with high-quality (symmetrical) output beams and by applying near-field focusing (discussed below).

Except in the drilling of very thin materials, taper cannot be eliminated. This problem can be minimised by making the proper choice of focal length and, in some cases, by refocusing between pulses for holes requiring several pulses for full penetration.

Holes ranging from less than 0.025mm in diameter up to around 1.5mm in diameter are routinely drilled by lasers, with depths reaching 25mm. For holes of over 1.5mm in diameter, a method called trepanning, which is actually cutting, is used. There are several ways in which this can be accomplished. In the simplest case, the part is rotated by an X–Y table or special fixturing about an axis which is not centred with the hole. Depending on the pulse rate of the laser being used, the speed of rotation is adjusted so that the holes drilled by the laser overlap. The percentage of overlap depends on the edge smoothness required. A figure of 50% is a reasonable overlap.

The rotation can also be accomplished by placing an optical wedge in the beam and rotating it, or by bringing the beam through a focusing lens off-centre and rotating the lens. These latter techniques limit the size of the hole that can be trepanned.

Fig. 7.1 illustrates a laser-drilled hole in metal, assuming that only one pulse has been used. The recast layer is molten material that was not ejected and solidified on the wall. Dross is solid matter not totally ejected from the bottom of the hole. Unlike the recast layer, dross is frequently loose and can be removed by brushing or light machining. In many

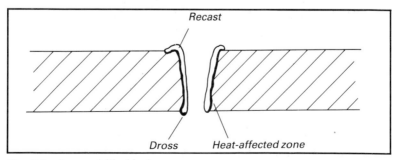

Fig. 7.1   Laser-drilled hole

applications the dross is negligible. The heat-affected zone (HAZ) is a layer which is formed where the microstructure has been altered by heat. The recast and HAZ are usually small compared with the hole size, and are usually only a few hundredths of a millimetre thick. Dross can be eliminated or minimised by the use of coatings or an air blast on the under-surface. Recast and HAZ thickness are minimised by optimising the drilling process, and their quantities can be reduced by using shorter pulses.

The most common lasers for drilling metals are Nd-YAG and Nd-glass. Typical pulse lengths for these lasers range from 0.1 to 3ms. CO_2 lasers are used to drill polymeric and ceramic materials. Pulse lengths ranging from 0.5ms to continuous wave (CW) are used. Pulsed operation of CO_2 lasers is preferred when processing ceramics, to reduce thermal stress cracking in the walls of the holes. Excimer lasers are also finding applications in the processing of non-metals (Fig. 7.2).

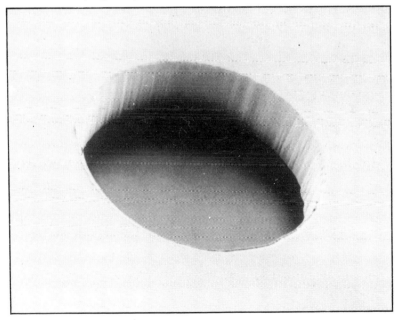

Fig. 7.2   300μm hole drilled in polyimide using a 0.248μm excimer laser (Courtesy: Lumonics Inc.)

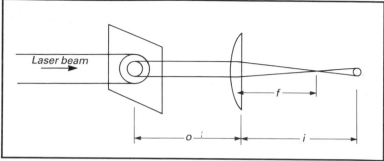

Fig. 7.3 Near-field focusing

The beam quality of high-power Nd-YAG and Nd-glass lasers is not very good, as a result of multi-mode operation. To improve focused spot quality, a technique called near-field focusing* is frequently used. In this technique, the beam is incident on an aperture which is slightly smaller than the beam. This truncates the beam and causes some power loss, but produces a highly symmetrical beam, although there will still be variations in power across the beam. The aperture is then imaged by a lens (Fig. 7.3).

According to the thin lens equation:

$$\frac{1}{o} + \frac{1}{i} = \frac{1}{f} \tag{7.1}$$

where o is object distance, i is image distance and f is lens focal length. Therefore:

$$i = \frac{fo}{o - f} \tag{7.2}$$

The size of the imaged spot is determined from the magnification:

$$m = -\frac{i}{o} \tag{7.3}$$

---

* The focusing technique previously described, in Chapter Two, is referred to as far-field focusing

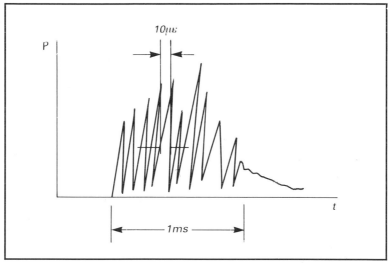

Fig. 7.4   Power output as a function of time for a solid laser

where the minus sign is meaningless for this purpose. Combining Eqns. (7.2) and (7.3) gives:

$$m = \frac{f}{o - f} = \frac{1}{o/f - 1} \tag{7.4}$$

Eqn. (7.4) makes it clear that to get a small spot, either f must be very small or o very large, so that the laser has to be placed a long distance away from the lens. This does not usually present a serious difficulty.

A major reason for the ability of solid lasers to pierce metals, ceramics and other materials readily is the fact that the pulses from these lasers are composed of many very short pulses (Fig. 7.4). This phenomenon is referred to as *spiking* and is the result of mode hopping. Each spike corresponds to the radiation from a group of ions which can radiate at a given mode of the oscillator. The population inversion is quickly depleted by stimulated emission, and another mode that has not been depleted starts to oscillate.

One of the great advantages of laser drilling is that it is possible to drill holes at large angles to the perpendicular to the surface: for example, holes can be drilled in metals at angles of over 60° to the perpendicular.

It is relatively simple to estimate the energy required to pierce holes, based on the energy balance approach. The energy required to drill a hole of radius a and depth l is given by:

$$V = Q\pi a^2 l \tag{7.5}$$

if Q, the energy required per unit volume to vaporise the material, is known. Q is also called the specific energy, and if unknown, may be calculated from $Q = \rho(Cp\triangle T_v + L_f + L_v)$, where $\triangle T_v = T_v - T_a$ (vaporisation temperature minus ambient temperature), ρ is density, Cp is specific heat at constant temperature, L_f is the latent heat of fusion and L_v is the latent heat of vaporisation.

Cutting

Cutting with lasers is a gas-assisted process (Fig. 7.5). The purpose of the gas-assist is <u>manifold.</u> Molten material is blown out of the kerf, preventing spatter from reaching the

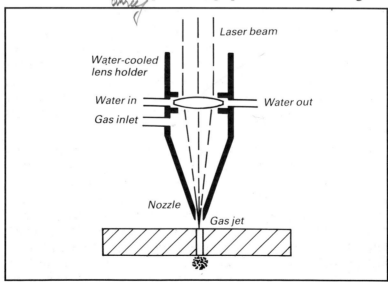

Fig. 7.5 Laser gas-assist cutting

lens or focusing mirror. Since the molten material is blown away, the amount of vaporisation required is minimised, thereby increasing both the cutting rate and the quality of the cut. This is due to the fact that there is less heat input than in pure vaporisation.

In cases where oxidisable material is being cut, oxygen or air may be used to enhance cutting rate and quality. Highly oxidisable metals such as titanium and aluminium should be cut with inert gases to prevent self-burning.

The use of nozzles

Nozzle design is a subject well beyond the scope of this book. However, it should be pointed out that the proper choice of nozzle and cutting parameters is critical to achieving quality cuts, especially in metals. The pressure in the nozzle may be as low as one-third of an atmosphere or as high as several atmospheres: good cuts can be obtained either way. The main advantages of high pressure are that it enables both faster cutting, and cutting of thicker materials (although this may not justify the additional gas consumption unless compressed air is used), and that improved quality of edge smoothness can be achieved.

The use of high-pressure nozzles is an attempt to achieve supersonic gas flow. Most nozzles cannot achieve this, regardless of the pressure, but it is possible to attain speeds close to Mach 1. With true supersonic flow, greater care must be taken to optimise the position of the workpiece relative to the nozzle, due to pressure variations in the gas stream.

In conventional low-pressure nozzles, the exit-hole diameter is about 1.5mm, and the nozzle-to-part standoff is somewhere between 1mm and 2mm. Control of focal-point position (optimally at the surface for metals) and nozzle standoff distance is critical. For cutting over large areas, even of flat stock, some passive or automatic feedback control system must be incorporated to maintain these dimensions within about one-quarter of a millimetre. Figs. 7.6 (a) and (b) show a laser-cut part and a close-up view of the cut edge.

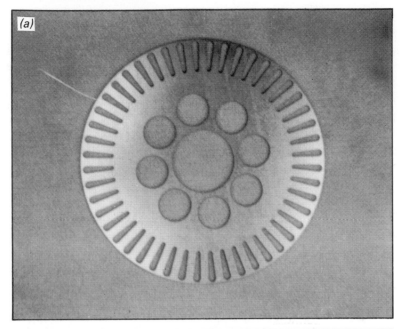

Fig. 7.6 (a) Laser-cut electric motor lamination part (Courtesy: Lumonics Material Processing Corp.), and (b) laser-cut low-carbon steel, showing edge

Requirements for the cutting process

Metals and polymers can be cut with CW CO_2 lasers, but for metals, better quality – i.e. smaller HAZ and smoother edges – is achieved by using pulsed operation. This is required for ceramics, and generally, CO_2 lasers are used for ceramics cutting, drilling and scribing, although Nd-YAG lasers can also be used.

Cutting with Nd-YAG lasers is always a pulsed operation, so that it is, in fact, a process of overlapping hole-piercing. Nd-YAG is very useful for cutting metals, including copper and brass alloys, as well as high-temperature materials used in aircraft engines, such as Hastelloy-X. Nd-YAG lasers can be used to cut metals up to 10mm thick, but most applications involve much thinner materials.

Power levels for CO_2 cutting lasers range from tens of watts for cutting thin plastics and ceramics, to over 1000W for cutting metals. Mild steel over 3cm thick can be cut with a CO_2 laser, but most industrial applications involve metals of 5–6mm thickness or less.

The edge smoothness that can be achieved in laser cutting is good. Edge surface finishes of less than 100μm RMS can be achieved. Heat-affected zones are typically less than 25μm.

A symmetrical beam spot shape is essential for cutting. To achieve a small focused spot size, a low-order mode is needed, preferably Gaussian. Near-Gaussian beam operation for CO_2 lasers of up to 2–3kW is not unusual, but high-power Nd-YAG lasers operate in multi-mode. The near-field focusing used with these lasers ameliorates the problem to some extent, but not entirely. Two- and three-dimensional cutting is done on multi-axis CNC and robotic systems, some of which are discussed in Chapter Ten.

Many modern CO_2 lasers provide a linearly polarised output which may be stable (fixed orientation) or not, depending on the laser design. Because of the variation of reflectance with angle of incidence and polarisation state, cutting quality is seriously affected by the polarisation state of the laser. In contour cutting, optimum speed and kerf can

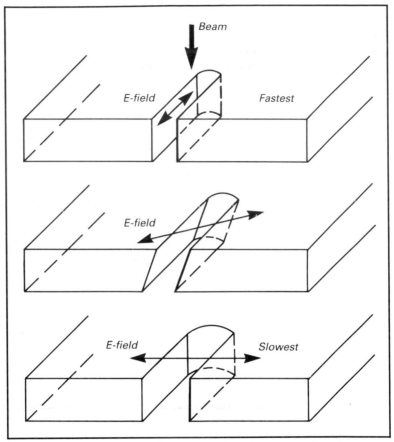

Fig. 7.7   Relationship of polarisation direction to metal cutting

vary by more than 50% for a linearly polarised laser, depending on how the polarisation direction relates to the cut direction. Fig. 7.7 illustrates this phenomenon.

A parallel arrangement is the ideal, a perpendicular one is worst and angles in-between result in sloping sides. Round holes will not be cut round and cut-out plugs may not drop out. The way to avoid this problem is to introduce a circular polariser into the beam path. As far as cutting is concerned, circularly polarised light has the same effect as unpolarised light. The laser must have a stable (fixed) linear polarisation,

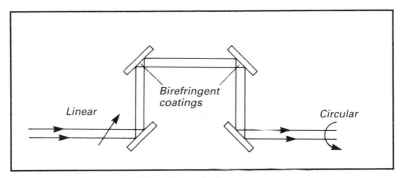

Fig. 7.8 Schematic of circular polariser

however, since the orientation of the direction of polarisation relative to the circular polariser is critical. Fig. 7.8 is an illustration of a circular polariser, also known as an enhanced cutting accessory.

Broad estimates of the power required for cutting can be made using a modified form of Eqn. (7.5), where:

$$P = Qwlv \qquad (7.6)$$

where w is the kerf width and v is the cutting speed. For example, plexiglas has a specific energy of about $8kJ/cm^3$. The power required to cut a 0.1mm-wide kerf in 6.35mm-thick plexiglas at 2.12cm/s is 266W. This is a fairly reasonable estimate. Due to the use of air as a gas-assist, less power or a higher speed could probably be used.

Marking

One of the major uses for lasers is in marking (sometimes referred to as engraving). The types of lasers used for such work are CO_2 and Nd-YAG. The CO_2 lasers are usually TEA lasers, since these are capable of producing rapid, short pulses of tens of Joules per pulse. The Nd-YAG lasers generally have an average output power of around 50W, and are Q-switched to give 100–150ns pulses, with peak powers of up to 60kW and repetition rates of up to 10kHz.

CO_2 lasers are used to produce entire patterns, barcodes, serial numbers, logos, etc., with one or two pulses, by

Fig. 7.9   Laser-engraved parts: left, wood engraved with CO_2 laser; right, anodised aluminium engraved with Nd-YAG laser

imaging a shadow mask onto the part. Because of the short pulse lengths, parts can be marked 'on the fly'. CO_2 laser marking is a simple and relatively inexpensive process. Some systems are capable of indexing for serialisation, but flexibility is limited. Fig. 7.9 shows laser-engraved anodised aluminium, and a CO_2 laser engraving. CO_2 lasers are used primarily for marking non-metallic substances, but when deep penetration marking is required, for example on engine blocks, high-power CO_2 lasers may be used.

Nd-YAG lasers are used for marking metallic parts. Typical engraving marks are only a few hundredths of a millimetre deep. The marking is accomplished by manipulating the beam with computer-controlled galvanometer mirrors and a shutter to actually 'write' the pattern. Writing speeds of 50cm/s are common, and patterns can be marked on areas of over 5 × 5cm by using a flat-field lens. The lens is protected from smoke and spatter by a disposable glass plate.

Chapter Eight

Alignment, gauging, ranging and triangulation

THIS chapter examines several laser-based optical inspection techniques used in industry. These are discussed using a minimum number of mathematical equa-
tions. The associated electronic and computing devices used
to interpret the optical information obtained are mentioned
but not discussed in depth.

Alignment

Alignment is perhaps the most obvious low-power laser
application. A laser beam travels in a straight line through
any transparent gas or liquid which has a constant refractive
index. The laser beam can be used in much the same way
that a string or chalk line is used for alignment in the construction industry. Tunnels, drainage tile and pipelines can
be aligned using a laser beam, and it has become common
practice to align industrial equipment using low-power laser
alignment systems.

Laser-based alignment systems generally consist of three
basic units. These units include a transmitter which produces laser light and structures it, a receiver, and a display
unit and/or computer. This section examines the basic principles of the optical components used in these systems.

Transmitters

HeNe and diode lasers are the most commonly used light sources in laser-based alignment systems. The HeNe laser is larger in physical size than the diode, but requires a smaller number of optical elements to structure the light it produces. A beam expander is frequently used to increase the HeNe laser beam diameter in order to reduce its divergence and irradiance (see Chapter Two, 'Optical processes'). An expanded beam maintains the same diameter over long distances, and the potential hazard to the human eye is reduced.

The diode laser is a much smaller device, but requires additional optics to structure its beam. Diode laser light is generated at the junction of a diode fabricated from a semi-conductor material such as gallium arsenide. The diode laser output window, or aperture, resembles a rectangular slit which is longer than it is wide. Due to this rectangular aperture, the laser beam diverges, and its far-field cross-section is elliptical (Fig. 8.1). A converging lens can be used to collimate the diode laser beam, but its cross-section will still be elliptical (Fig. 8.2). Cylindrical lenses or a prism pair can be used to circularise the laser beam cross-section. Fig. 8.2 also shows a prism pair being used to circularise the laser beam by expanding the elliptical beam parallel to its minor

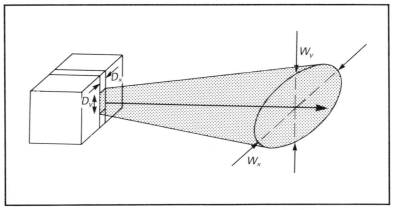

Fig. 8.1 Diverging laser beam from rectangular diode laser aperture ($D_x < D_y$ and $W_x > W_y$)

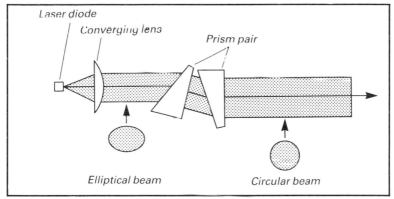

Fig. 8.2   Diode laser optics

axis. The divergence and irradiance of this beam can be further reduced by adding a beam expander.

In some applications it is desirable to scan the laser beam to sweep out a flat plane. This can be accomplished by using a rotating mirror or penta-prism. Laser beam scanning is discussed in further detail in the section on 'Laser beam scanner systems' later in this chapter.

Detectors

The human eye can be used in numerous applications to observe a HeNe laser beam incident on an object or target that is to be aligned. However, the diode laser light which is commonly used in alignment systems is invisible to the human eye. When automation or greater accuracy is requi red, laser beam position detectors can be used. A variety of silicon photodiode position detectors are available in one- or two-dimensional formats.

A simplified diagram of a laser alignment system with a quadrant detector is shown in Fig. 8.3. Each quadrant of the detector is a separate photodiode which produces an electrical output proportional to the light power it receives. If the incident laser beam is centred on the detector, each segment of the quadrant detector receives the same amount of power. When the laser beam is not centred, one or two quadrants of the detector will receive more light power.

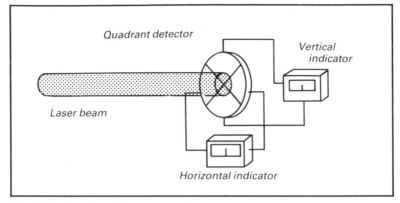

Fig. 8.3   Quadrant detector system

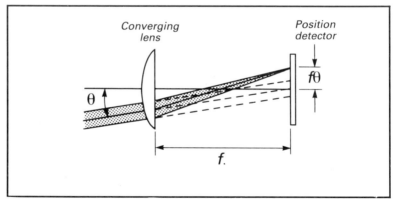

Fig. 8.4   Angular alignment using a converging lens

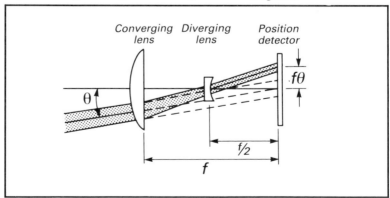

Fig. 8.5   Angular alignment using a confocal lens pair

Systems have been designed which use the outputs from quadrant detectors to give the laser beam position relative to the detector centre. Recent advances in computer vision systems have made two-dimensional diode array detector systems widely available in industry. The resolution of these devices is dependent on the spacing and/or the number of photodiodes. For one-dimensional centring, linear photodiodes or lateral effect photodiodes can be used.

In many applications angular alignment is also necessary. Angular alignment can be measured by using a converging lens, as illustrated in Fig. 8.4. If the laser beam is originally centred on the detector without the lens, the focused beam will be shifted by f_θ at the detector when the lens of focal length f is inserted. This displacement can be used to determine the angular alignment θ. A focused beam can be used with lateral-effect or diode array detectors. Quadrant detectors require a larger diameter beam at the detector so that all four quadrants of the detector receive light power. This can be accomplished by using a converging lens combined with a diverging lens, as illustrated in Fig. 8.5. With this confocal lens pair the beam displacement is still f_θ, but the beam diameter is reduced by one-half rather than being focused to a point.

Laser-based alignment systems designed to determine both the position and the angular alignment of a laser beam are available commercially. One application for such systems is spindle alignment. A spindle is a slender rod that can be used to hold and rotate a machine tool such as a drill. To drill a hole in the proper position at the correct angle, the spindle must be aligned relative to a standard. To accomplish this, a laser transmitter can be mounted on the spindle, and the laser receiver placed in the holes of a standard part. Fig. 8.6 shows a laser alignment system designed for spindle alignment.

Distance measurement

One of the first proposed laser applications was to find the distance between the Earth and the Moon. This experiment was made possible when retroreflectors were placed on the

Moon during one of the 'Apollo' missions. A pulsed laser was used to determine the time required for light to make a round trip from the Earth to the retroreflectors and back. One-half of this time interval was then multiplied by the speed of light, to determine the distance between the Earth and the Moon. This time-of-flight technique was accurate to within one metre. Other laser-based distance measurement techniques have been developed which are more suitable for industrial measurements. These techniques use continuous wave lasers and include amplitude modulation, triangulation and interferometry. Interferometry techniques can be used to measure displacements accurately to within a fraction of the wavelength of the light used to make the measurement. Amplitude modulation and triangulation techniques can be used to measure displacements with greater accuracy than the time-of-flight technique, but with less accuracy than interferometry techniques.

Fig. 8.6   Laser-based spindle alignment system with transmitter (A), receiver and lens system (B), computer (C), and translation stage (D) – used for system calibration

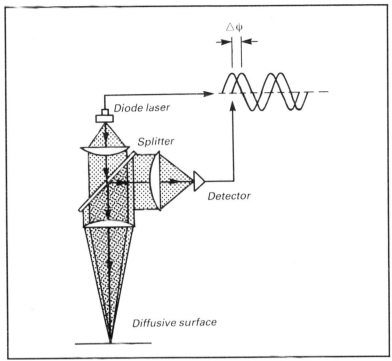

Fig. 8.7 Optical system designed to measure distance, using the phase shift of an amplitude modulated laser beam

Amplitude modulation

In this range-finding technique a laser beam is amplitude modulated by changing the laser beam power sinusoidally. This can be conveniently accomplished by sinusoidal variation of the input current to a diode laser. The modulation wavelength λ_M is c/f_M, where c is the speed of light and f_M the modulation frequency. As shown in Fig. 8.7, this modulated beam can be focused on a diffusive target surface by using a converging lens. The same lens can be used to focus some of the light scattered from the surface to a detector. To accomplish this, a beam splitter can be used to reflect scattered light to the detector while also allowing light from the diode laser to be transmitted through the splitter to a target. To determine the range, the phase of the diode laser current input can be compared to the phase of the detector output. The phase difference between these two signals is as a result of the amount of time required for the light to make the round trip from the transmitter to the target and back. Techniques for measuring the phase difference with extremely high accuracy have been developed. By restricting the range to be measured to values between $n\lambda_M/2$ and $(n + 1)\lambda_M/2$, where n is an integer $0, 1, 2, 3 \ldots$, the range can be measured without ambiguity. Range R can be determined using the equation:

$$R = \lambda_M/2 \left[(\triangle \phi/2\pi) + n \right] \qquad (8.1)$$

where $\triangle \phi$ is the phase difference.

This technique is a non-contact method which seems to have excellent potential. By scanning the modulated beam, it is possible to obtain information which can be used for three-dimensional computer vision.

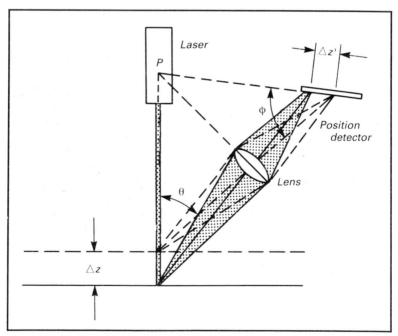

Fig. 8.8 Optical triangulation unit designed to make one-dimensional measurements

Triangulation

Two types of laser-based triangulation detectors are now being used in industry: systems which project a single laser light spot onto a surface to make a one-dimensional measurement, and systems which can make two-dimensional measurements by projecting a line of laser light.

Fig. 8.8 is a simplified diagram of an optical triangulation system designed to make one-dimensional displacement measurements in the direction parallel to the z-axis. A HeNe or diode laser is used to produce a spot of light on a diffusive surface. Scattered light from the surface is focused by a converging lens onto a linear position detector. The electrical output from this detector can be used to determine the laser spot image position on the detector. Diffusive surface displacement parallel to the incident laser beam causes a corresponding image displacement on the position detector. To maintain the best image focus on the position detector, the Scheimpflug principle can be used. According

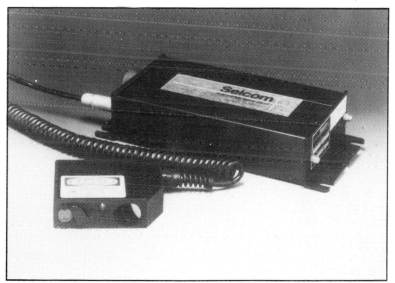

Fig. 8.9 *Optical triangulation system (Courtesy: Selective Electronic – Selcom)*

to this principle, the laser beam, an extended lens diameter, and a line passing through the detector all intersect at a point P (Fig. 8.8). By using plane geometry, the relationship between angles θ and ϕ can be determined to be:

$$m \tan\phi = \tan\theta \tag{8.2}$$

where m is the magnification. Magnification m is $-i/o$, where the image and object distances for the lens are i and o, respectively. Displacement $\triangle z'$ of the image on the position detector is given by:

$$\triangle z' = m\,(\sin\theta/\sin\phi)\,\triangle z \tag{8.3}$$

where $\triangle z$ is the diffusive surface displacement component parallel to the laser beam.

Laser-based triangulation systems of this kind are extensively used in industry. One application for such systems is to determine the level of molten metal in a pour box. By using an optical triangulation unit mounted on the arm of a robot, proper stand-off distance can be maintained so that the robot can perform operations like welding and painting. Two triangulation units can be used to determine the thickness of sheet stock or belting. Fig. 8.9 shows a triangulation unit, designed using the geometry discussed previously.

Two-dimensional measurements can be made by projecting a line of laser light onto a surface and using a two-dimensional detector. With the introduction of computer vision into industry, systems of this type are now available from several vendors. The system illustrated in Fig. 8.10 can make dimensional measurements in the x and z directions. To produce the laser light line on the part to be measured, a cylindrical lens is used. This lens expands the laser light in one direction, but not in the other. Again, the Scheimpflug principle can be used to place the detector at the proper angle to maintain the best laser light line image focus on the detector. This light line image is illustrated in Fig. 8.10. Distance $\triangle z'$ can be used to determine $\triangle z$ by using Eqn. (8.3), and distance $\triangle x'$ can be used to determine $\triangle x$ by using:

$$\triangle x' = m \triangle x \tag{8.4}$$

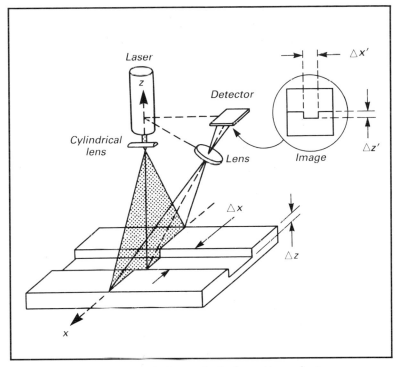

Fig. 8.10 Optical triangulation unit designed to make two-dimensional measurements

where *m* is the magnification. Note that this system is not designed to make measurements in the *y* direction. Triangulation devices of this type are used to make measurements of the depth and width of seams and gaps in sheet metal.

Laser interferometry

A coherent laser beam produces an electric field which can be thought of as a sinusoidal travelling wave (see Chapter One, 'Coherence'). If the electric fields of two interfering coherent laser beams are in phase, the electric fields add, resulting in an irradiance greater than the irradiance produced by either beam. If the electric fields are 180° out of phase, the two fields subtract, resulting in a lower irradiance.

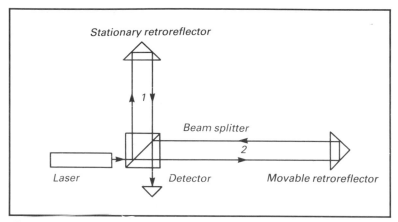

Fig. 8.11   Laser interferometer (simplified)

Fig. 8.11 shows a laser interferometer system frequently used in industry. A laser beam is split into two beams ((1) and (2)) by a beam splitter. Beam (1) is reflected to a stationary retroreflector, which reflects this beam back to the beam splitter. Beam (2) is transmitted through the splitter to a movable retroreflector, which also reflects this beam back to the splitter. At the splitter, a portion of beam (1) is transmitted, and a portion of beam (2) is reflected, thus superimposing the two beams. The superimposed beams are incident on a detector. If the optical path lengths travelled by the two beams are identical, the two beams will be in phase and will have maximum brightness at the detector. By translating the movable retroreflector a distance of one-quarter wavelength directly away from or towards the beam splitter, the optical path length of beam (2) is increased or decreased by one-half wavelength. Now the beams are 180° out of phase, and the light level at the detector is at a minimum. For each quarter-wavelength displacement, the light level will change from a high to a low value, or vice versa.

The system illustrated in Fig. 8.11 can measure displacement magnitude, but not direction. Commercial systems use a two-frequency laser, or circular polarisation techniques, to determine displacement direction as well as magnitude. Laser interferometers provide one of the most accurate

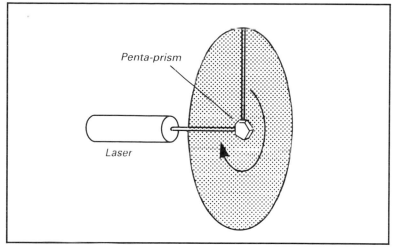

Fig. 8.12 Penta-prism scanner

measurement tools available to industry. These systems are so sensitive that temperature, humidity, and atmospheric pressure must be taken into account to obtain maximum accuracy. One quarter-wavelength for HeNe laser light is 0.158μm. It should be noted, however, that the movable retroreflector must be mounted to the object which incorporates a laser interferometer as an integral part, in order to make linear displacement measurements. Most often, laser interferometers are used to calibrate linear measurement devices. Techniques have also been devised, using this equipment, to check surface flatness.

Laser beam scanner systems

Levelling

A penta-prism can be used to turn a laser beam through a right angle even though the prism is not precisely aligned. By rotating the prism about the axis of the incident beam, as illustrated in Fig. 8.12, the output beam sweeps out a flat plane. Systems of this type are used on construction sites to align ceilings, walls, and foundations. In many cases a HeNe laser is used, so that the scanning laser beam produces a red

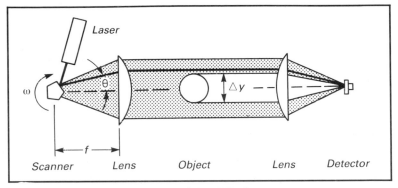

Fig. 8.13   Laser scanner gauge (simplified)

line on the object which can be viewed by a human observer in order to align the object. If greater accuracy is required, or automated alignment is desirable, a position detector system can be used. By using a laser beam which scans out a horizontal plane, an automated position detector system can be used to detect the beam and control the blade on earthmoving equipment, so that an absolutely flat surface can be graded. Farmers and road builders use these systems to grade their fields and road beds.

Gauging

A rotating mirror can be placed at the focal point of a converging lens to produce a scanning laser beam that remains parallel to the lens axis. Polygon mirrors are frequently used to obtain multiple scans per mirror rotation. Fig. 8.13 shows a laser scanner system which can be used as a micrometer. For small angles θ, the beam exiting the transmitter lens scans across a plane parallel to the lens axis at a speed of:

$$v = 2\omega f \tag{8.4}$$

where f is the lens focal length and ω is the angular velocity of the rotating mirror. An object placed in the scan plane will block the beam for a time period $\triangle t$.

The dimension y is then:

$$\triangle y = v \triangle t \tag{8.5}$$

or:

$$\triangle y = 2\omega f \triangle t \tag{8.6}$$

A second converging lens can be used to focus the scanning beam onto a photodiode. By measuring the time interval that the laser beam is blocked, the object dimension $\triangle y$ can be determined.

Several vendors sell systems based on this principle, which are interfaced with computers for data processing. Fig. 8.14 shows a laser scanner unit being used to measure the diameter of a cylindrical part. An interesting application of this device is to measure the diameter of wire or cylindrical shafts exiting a production line immediately after they are produced. By interfacing the scanner unit with the production equipment, feedback can be provided to compensate for process variations.

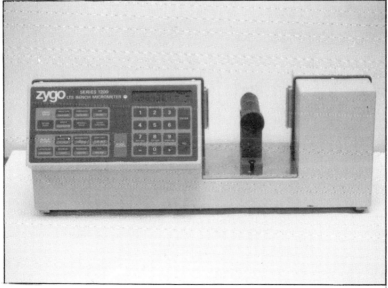

Fig. 8.14   Laser bench micrometer (Courtesy: Zygo)

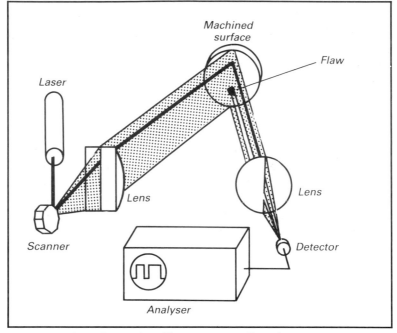

Fig. 8.15   One-dimensional laser scanner flaw detector

Surface inspection

Fig. 8.15 is a simplified diagram of a surface inspection system. In many systems, an oscillating or rotating mirror is placed at the focal point of a converging lens to produce a scanning laser beam which remains parallel to a lens axis. In Fig. 8.15 it is assumed that the machined surface is smooth, almost specular, except for surface flaws which tend to be filled with black residue from the machining process. The laser beam is scanned across the machined surface. As the light scans the specular surface, it is reflected to a second converging lens and focused onto a detector. When the beam scans a flaw, the light is absorbed, and the amount of specular reflected light decreases. Thus, in observing the reflected light with the detector, the light level will be high when the laser beam scans the smooth machined surface and low when it scans a flaw. The signal output from the detector can be evaluated by using a computer.

Another technique that can be used is to observe the light scattered from a surface. A detector placed to observe scattered light will detect very little light as the laser beam scans the specular surface, but will detect more light as the beam scans a flaw. This technique is particularly good for detecting scratches on an otherwise specular, mirror-like surface.

To inspect entire surfaces, one technique is to move the object being inspected so that the entire surface is scanned. Another technique is to scan the laser beam in a raster pattern by using two rotating or oscillating mirrors. A two-mirror scanner system is shown in Fig. 8.16. Laser light is focused on the surface by a converging lens. One mirror rotates or oscillates rapidly to scan the beam in the horizontal direction, while the other rotates or oscillates more slowly to move the scan line vertically. In the system illustrated in Fig. 8.16 reflected light is received back through the converging lens and is then reflected by a beam splitter to a detector.

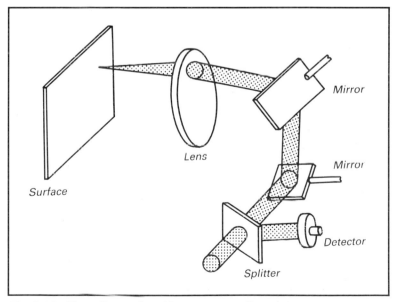

Fig. 8.16   Two-dimensional laser scanner flaw detector

Barcode scanners

One solution to inventory control in industry is to use barcode scanners to identify, track, locate and count the industrial components being produced, shipped or received. In this application a barcode is inscribed on or attached to each industrial component. Usually the bars are black and the spaces in between the bars are white. As a laser beam scans across the bars, it is reflected by the white surface and absorbed by the black surface. The reflected light is focused on a detector by a system such as that shown in Fig. 8.16. The light level at the detector (and therefore the electrical output from it) will alternate between a high and a low value as the laser beam scans the barcode. A computer system interfaced with the detector is used to measure the relative widths of the black and/or white bars. If a binary code is used, the black bars, for example, will have only two widths. One width bar is interpreted as '1' and the other as '0'.

Since relative widths are measured, the laser beam can scan the bars at any angle, provided that all the bars are scanned. Thus, many barcode scanners scan the beam through a complex 'criss-cross' pattern to ensure that the entire barcode is read by the system. A pair of bars is used to represent a number between 1 and 10. Barcodes for base 10 numbers thus consist of a series of bar pairs. Many items sold have a printed barcode on the item or container which can be interpreted as two five-digit numbers by a barcode scanner system. An alternative to using mirrors to scan a laser beam is using a holographic optical element. Holography is discussed in the next chapter.

Chapter Nine

Laser velocimetry and holographic and speckle interferometry

ALL THE techniques discussed in this chapter are dependent on the interference of coherent laser light. To date, the primary applications of these techniques have been limited to the research laboratory. However, some applications are now appearing in the field and on the industrial production line. All these optical techniques are non-contacting and have excellent potential for non-destructive testing.

Velocimetry

Velocity measurements using laser velocimetry can be explained using two theories: the interference of light and the Doppler effect. These theories are used below to explain the operating principles of a laser anemometer, often known as a laser Doppler velocimeter (LDV) or a laser Doppler anemometer (LDA). The Doppler effect is then used to explain the basic operating principles of a laser Doppler vibration detector.

Laser anemometry

Measuring particle velocity using laser anemometry is standard practice for engineers and scientists studying gas and fluid flow. When two coherent laser beams which have the

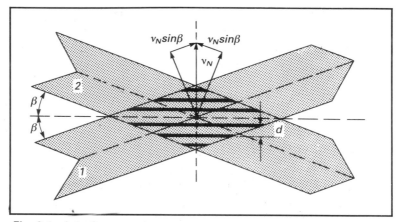

Fig. 9.1 *Interference planes resulting from two intersecting coherent beams*

same plane of polarisation intersect, bright and dark interference planes are formed in the region in which the beams overlap (Fig. 9.1). These interference planes are parallel to a plane bisecting the angle between the two beams. Adjacent bright or dark planes are separated by the distance, d, given by:

$$d = \lambda/(2 \sin\beta) \qquad (9.1)$$

where λ is the laser light wavelength and 2β is the angle between the two intersecting laser beams. It is assumed that small particles are contained in a transparent gas or liquid, and travel at the same velocity as the medium they are in. Thus, measuring particle velocity is tantamount to measuring flow velocity. In some gases or fluids, these particles already exist; for example, the carbon particles in the exhaust gas from a diesel engine. Otherwise, the gas or liquid must be 'seeded' with particles. As a particle passes through the interference planes, it scatters light when travelling through bright planes, but not when passing through dark planes. This scattered light will 'blink' at a frequency, f, given by:

$$f = 2(v_N/\lambda) \sin\beta \qquad (9.2)$$

where v_N is the magnitude of the particle velocity, which is assumed to be perpendicular to the interference planes.

Thus, the velocity of a particle can be determined by simply measuring the 'blink' frequency.

Using the Doppler effect, Eqns. (1) and (2) can be verified, and a technique for determining velocity direction explained. Assuming that light from a stationary laser is reflected from an object moving with a velocity v directly toward or away from a laser, this reflected light experiences a frequency shift, δf, given by:

$$\delta f = 2f_0 v/c \qquad (9.3)$$

where f_0 is the laser light frequency and c is the speed of light. The velocity is positive for motion toward the laser, and negative for motion away from the laser. For a particle with a velocity v_N, as shown in Fig. 9.1, the particle has a velocity component $v_N \sin\beta$ when travelling in the *same* direction as laser beam (1) and when travelling in the *opposite* direction to laser beam (2). The reflected light from beam (2) will have a higher frequency than the reflected light from beam (1). By mixing the light produced by both beams, the mixed light will 'beat', or 'blink', as described in Chapter One, 'Coherence'. The beat frequency is the difference between the reflected light frequencies of beams (1) and (2). This is given by Eqn. (9.2). By changing the frequency of either beam (1) or (2) with a device such as a Bragg cell, the direction of v_N can be determined. Assuming the geometry and the velocity direction indicated in Fig. 9.1, the beat frequency will *increase* if the frequency of beam (2) is increased. If the velocity is in the opposite direction, the beat

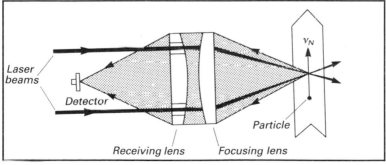

Fig. 9.2   Laser Doppler velocimeter

frequency will *decrease* when the frequency of beam (2) is increased.

In gas or fluid flow experiments, all velocities cannot be expected to be conveniently perpendicular to the interference planes produced by intersecting laser beams. For a particle with random velocity direction, the system described here is only sensitive to v_N – the velocity component perpendicular to the interference planes.

Other velocity components are usually found by producing orthogonal sets of interference planes with different-coloured laser beams in the same region. Reflected light is frequency-filtered to separate the colours, which can then be directed to individual detectors. Fig. 9.2 shows a laser Doppler velocimeter which can be used to determine a velocity component of a particle.

Laser Doppler vibrometry

Fig. 9.3 shows a device that can be used to measure vibrating surface velocity by evaluating the Doppler frequency shift of light reflected from a surface. Laser light is focussed onto a surface, using a converging lens, at an angle β relative to the lens axis. That portion of the light which is reflected back to the lens at angle β is reflected by a mirror to beam splitter (2). The phase shift of this reflected light is given by:

$$\delta f = 2f_0 \, (v_F/c) \cos\beta \qquad (9.4)$$

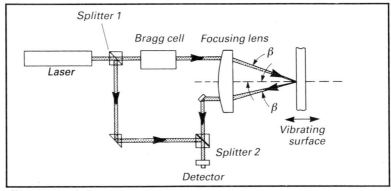

Fig. 9.3 *Laser Doppler vibration detector (simplified)*

where v_p is the surface velocity component parallel to the lens axis. Beam splitter (1) is used to reflect a portion of the incident laser beam to a mirror which also directs this beam to beam splitter (2). Beam splitter (2) is used to mix the reflected light with light coming directly from the laser. Being of two different frequencies, the mixed light beats at a frequency of δf. This process is called heterodyning. Heterodyned light is directed onto a detector which produces a voltage output to determine the component of the surface velocity parallel to the lens axis. By frequency-shifting the incident light with a Bragg cell, the velocity component direction can also be determined. By integrating the signal with respect to time, it can be used to determine displacement along the direction of the lens axis, and by differentiating the signal, it can be used to determine acceleration.

Holographic interferometry

A hologram can be thought of as an optical device that is capable of producing three-dimensional images. In industry, one of the most important applications is holographic interferometry. This can be used to measure surface deformations with accuracies of the order of one-quarter wavelength of the laser light used to record the holograms. This section explains how to record and reconstruct holographic images. There then follow a general discussion of holography, and examinations of real-time, double-exposure and time-average holography. These conventional holographic techniques are sensitive to surface displacement magnitude. A carrier fringe technique is described which allows displacement direction to be determined.

Many industries design prototype parts by using computer-based finite element analysis. When a prototype part has been designed and built, it is usually tested to verify the finite element analysis or to determine how it vibrates or deforms in a real-life application. Holographic interferometry, when properly used, can perform these tests quite effectively.

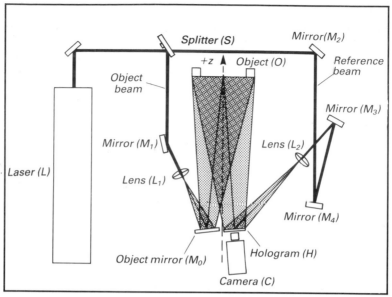

Fig. 9.4   Holography system geometry, designed for simplified interpretation of interference fringes due to surface displacement

Fig. 9.5   The holography system illustrated in Fig. 9.4

Recording and reconstruction

Figs. 9.4 and 9.5 show a holography system. Honeycomb steel or granite tables supported by air tubes are usually used to provide a vibration-free surface for the optical components used in holography. Lasers commonly used in holography are HeNe, argon, krypton and pulsed ruby. HeNe, argon and krypton lasers can be used to evaluate steady-state phenomena, and the ruby laser can be used to evaluate transient phenomena. The laser beam is first split into two beams, called the reference and the object beams (Fig. 9.4). The object beam is spread by a short focal-length lens and is then reflected by the object beam mirror so that it can be used to illuminate an object uniformly.

To obtain three-dimensional images, the object surface should be diffusive and is often painted with flat white paint to accomplish this. Light scattered from the diffusive surface is incident on a photosensitive recording medium, such as a photographic plate. Reference beam light is reflected by mirrors and spread by a lens to illuminate the recording medium uniformly. The light scattered from the object and the reference beam light interfere at the recording medium. After proper exposure to these interfering beams, the recording medium is developed and becomes a hologram.

The hologram can be 'played back', or reconstructed, by using a beam identical to the reference beam used to record the hologram. By viewing the light diffracted by the recorded interference fringes (Fig. 9.6), the observer sees a three-dimensional image of the object in the same position it occupied when the hologram was recorded.

Real-time holographic interferometry

In real-time holographic interferometry, a hologram of an object is first made. It is convenient to record this hologram on a recording medium such as thermoplastic, so that the hologram can be exposed and developed in situ. With the developed hologram placed in the same position as it was when exposed, it is played back with the reference beam, and the object beam is used to illuminate the object. An observer then views the object through the hologram. The

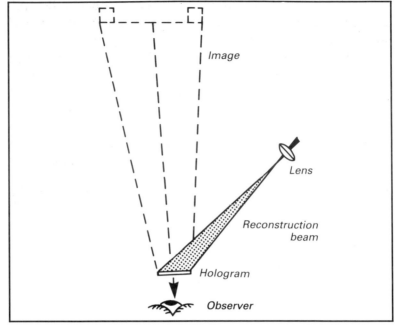

Fig. 9.6   Holographic reconstruction

observer sees the illuminated object and the holographic image simultaneously. If the process is properly executed, the holographic image and object will be in the same position. The optical path-length of the object beam is then changed by deforming the object or by changing the refractive index of the transparent medium through which the object beam passes. Due to optical path-length changes, the light reflected from the object and the light producing the holographic image will interfere, producing bright and dark fringes. As seen by the observer, these fringes normally appear on the surface of the object being deformed. For a positive recording medium, such as thermoplastic, bright fringes are formed when the optical path-length of the object beam changes by an integral number times the laser-light wavelength. This is due to the constructive interference of the light reflected from the object and the light diffracted by the hologram. Dark fringes are observed when the light reflected from the object and the light diffracted from the hologram destructively interfere. This occurs when the optical path length of the object beam changes by an integral number times the wavelength plus one-half wavelength.

The holography system described in Fig. 9.4 is one that can be used to determine surface displacement components in the z direction. To simplify the mathematics required to determine displacement, the system geometry meets certain requirements:

- The z-axis bisects the angle between the object beam light that illuminates the object and the object beam light that is scattered back to the hologram.
- The angle between the object beam light illuminating the object and the light scattered back to the hologram is small, or equal to zero.

Under these conditions, the change in the object beam optical path-length can be assumed to be approximately twice the surface displacement component in the z direction. For bright fringes:

$$d_z = m\,\lambda/2 \tag{9.5}$$

where λ is the light wavelength, d_z the displacement component in the z direction and $m = 0, 1, 2, 3, \ldots$. Fig. 9.7 shows a circular diaphragm which was clamped

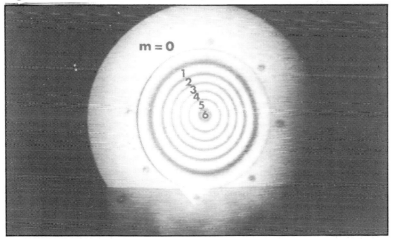

Fig. 9.7 Real-time holographic image with fringes due to normal displacement of diaphragm centre

around its outer perimeter and pushed forward, $-z$ direction, at its centre. The clamped region did not move and is represented by $m = 0$. In the bright circular region, represented by $m = 1$, the diaphragm moved forward one-half wavelength. If a HeNe laser was used, $\lambda = 0.633\mu m$, and this first bright fringe corresponds to a displacement in the $-z$ direction of $0.317\mu m$. Counting the fringes inwards from the clamped region, the diaphragm centre is represented by $m = 6$. Thus, it moved forward a distance of $1.90\mu m$.

Dark fringes can be represented by $m = \frac{1}{2}, \frac{3}{2}, \frac{5}{2}, \ldots$ For the holography system discussed, Table 9.1 lists displacement magnitudes in terms of m and the wavelength λ. Without previous knowledge, the observer cannot determine the displacement direction from the fringe pattern shown in Fig. 9.7. The fringe pattern would look the same if the diaphragm was pushed in the $+z$ direction to a distance of $1.90\mu m$. A technique to determine the displacement direction is discussed in the section on 'Carrier fringes' later in this chapter.

Two disadvantages of real-time holographic interferometry are that it requires a continuous wave laser and that once the hologram has been removed from the holography system, it cannot be used by itself to reconstruct interference fringes. The observer must view the object through the hologram so that the holographic image and the

Table 9.1   List of displacement magnitudes and vibration amplitudes for low-order fringes, in terms of laser light wavelength

Fringe order	Fringe type	Displacement magnitude (λ) (real-time and double-exposure)	Amplitude (λ) (time average)
0	Bright	0	0
0.5	Dark	0.25	0.19
1.0	Bright	0.50	0.30
1.5	Dark	0.75	0.44
2.0	Bright	1.00	0.56
2.5	Dark	1.25	0.69
3.0	Bright	1.50	0.81
3.5	Dark	1.75	0.94
4.0	Bright	2.00	1.06
4.5	Dark	2.25	1.19

original object are superimposed. It is advantageous to observe the superimposed interference fringes, the holographic image and the original object with a video camera and a zoom lens. By videotaping these images, a permanent record can be obtained for data analysis and documentation. Another technique that can be used to obtain a permanent record of interference fringes and that permits the use of a pulsed laser is double-exposure holographic interferometry.

Double-exposure holographic interferometry

Double-exposure holography can also be used to determine surface displacement magnitudes. This holographic technique can be explained using Fig. 9.4. In double-exposure interferometry, the hologram is first exposed with the object in an undeformed state. After the first exposure, the object is deformed. Finally, the hologram is exposed a second time and developed. The developed hologram can be played back using the original reference beam or its equivalent. When the hologram is viewed, the observer sees two superimposed holographic images. Interference fringes are formed as a result of the interference from the light which is producing the two holographic images. These interference fringes are identical to those formed in real-time holographic interferometry. Surface displacement magnitude in the z direction is determined as before by using Eqn. (9.4) or Table 9.1.

A double-exposure hologram can be stored, and the interference fringes viewed at any time, by playing the hologram back with a reconstruction beam equivalent to the original reference beam used to produce the hologram. Thus the hologram itself can be used as a permanent record of the investigation. Double-exposure holography is the primary technique employed to evaluate surface displacement by holographers which use photographic film or a pulsed ruby laser. The disadvantage of this technique is that the observer cannot see the interference fringes form as the object is deformed. Fig. 9.8 shows a double-exposure image of an operating engine, produced by using a pulsed ruby laser and played back with a HeNe laser. The fringe pattern was used to evaluate surface displacement resulting from engine vibration.

Time-average holographic interferometry

The alternative to using a pulsed ruby laser for vibration analysis is to use a continuous wave (CW) laser and time-average holographic interferometry. In this technique, the object is usually forced to vibrate in resonance with an acoustic or mechanical vibration source while the hologram is exposed.

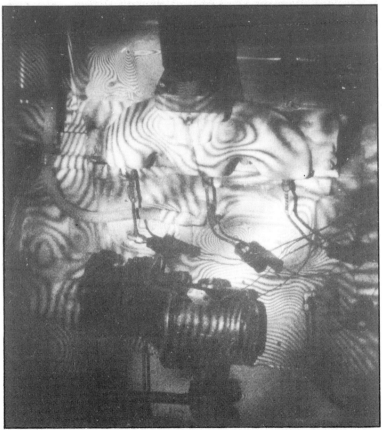

Fig. 9.8   Double-exposure hologram image of an operating engine, showing fringes resulting from vibration

However, in order to use this time-average technique, the object must vibrate with a constant amplitude and mode shape. For the purposes of this explanation, it is assumed that the holography system is configured as shown in Fig. 9.4, and that the object is a circular diaphragm.* When the diaphragm vibrates in its first mode of vibration, its perimeter is stationary and its centre region vibrates in the z direction with simple harmonic motion. Fig. 9.9(a) shows the image produced by a time-average hologram of the diaphragm vibrating in its first mode. For each surface segment vibrating in simple harmonic motion, the surface will have zero velocity when the diaphragm displacement has its maximum or minimum values. Surface velocity due to the vibration is at its maximum at a position midway between the diaphragm displacement maximum and minimum. Thus, each surface segment vibrating in simple harmonic motion spends more time at its maximum and minimum displacement positions and less time at any other position.

To simplify, the hologram can be thought of as receiving light only when the diaphragm is at the maximum and minimum displacement positions, and as not receiving light when the diaphragm is at other positions. The distance between the positions of maximum and minimum displacement is equal to twice the vibration amplitude. According to the holography system illustrated in Fig. 9.4, a dark interference fringe is formed if the distance between the maximum and minimum displacement is one-quarter wavelength, and a bright interference fringe is formed if this distance is one-half wavelength. Thus, a dark fringe is formed if the amplitude is one-eighth of a wavelength, and a bright fringe is formed if the amplitude is one-quarter wavelength. (This explanation is oversimplified, since the light reflected from the diaphragm is in fact received by the hologram for all positions between the maximum and minimum displacements. A more complete mathematical analysis indicates that bright *and* dark fringes are observed when the vibration amplitude is equal to those values given in Table 9.1. While

---

*The explanations provided here are oversimplified. The references listed in the bibliography can be used to obtain more extensive information

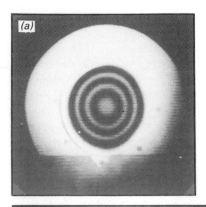
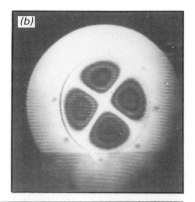
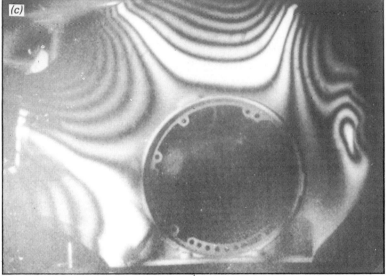

Fig. 9.9   Time-average holographic images: (a) circular diaphragm vibrating in first resonant mode, (b) circular diaphragm vibrating in higher order resonant mode, and (c) vibrating transmission case

this simplified explanation provides an intuitive way to think about time-average holographic interferometry, it does not yield an accurate value for the vibration amplitude.)

Fig. 9.9(b) shows a diaphragm vibrating in a higher order resonant mode. Fig. 9.9(c) shows the bell portion of an automotive transmission case vibrating at one of its resonant frequencies.

Carrier fringes

The real-time and double-exposure holographic interferometry techniques described previously can be used to determine displacement magnitudes, but not their direction. Carrier fringes provide a convenient way to determine displacement direction. To produce carrier fringes, the object mirror shown in Fig. 9.4 is rotated a small amount between exposures (in double-exposure holography), or after the hologram is developed (when using real-time holography). If the circular diaphragm used in previous examples is not displaced, rotation of the object mirror creates parallel fringes like those shown in Fig. 9.10(a). Diaphragm displacement causes carrier fringe deformation (Fig. 9.10(b)). The direction of this deformation depends on the direction of the diaphragm displacement. If the rotation direction of the mirror and the carrier fringe deformation direction are known, it is possible to determine displacement direction.

Speckle interferometry

When laser light reflected from a diffusive surface is viewed, laser speckle is immediately apparent. That is, light reflected from a diffusive surface will appear to be made up of closely spaced bright and dark speckle spots. This phenomenon is due to the constructive and destructive interference of light reflected from the small peaks and valleys on a

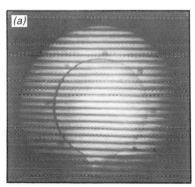 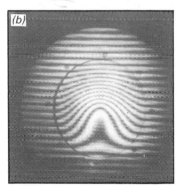

Fig. 9.10 Carrier fringes: (a) undeformed and (b) deformed diaphragm

diffusive surface. When this light is viewed by the human eye, it constructively and destructively interferes at the retina of the eye, creating the speckle pattern. If the observer views this light through a small aperture, the speckle size increases. This can be demonstrated by observing laser speckle through a small crack between two fingers. To photograph holographic images, a large lens aperture is often used in order to reduce the speckle spot size recorded on film.

Speckle interferometry can be used to determine surface displacement in much the same way as holography. Fig. 9.11 is a simplified sketch of one type of speckle interferometry system. One difference between the two techniques is that a high-resolution two-dimensional solid-state or tube-type image sensor is used, rather than a photosensitive recording medium such as a photographic plate. As in holography, the sensor is illuminated with light reflected from the object and with light coming directly from the laser. These beams are again known as the object and reference beams. The diffusive surface illuminated with laser light is imaged on the image sensor. Without the reference beam, the image is made up of speckle spots. Speckle spot size can be changed by changing the diameter of the camera aperture. With the

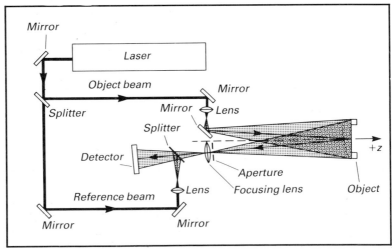

Fig. 9.11   Simplified speckle interferometry

reference beam present, light forming each speckle spot on the image interferes with the light from the reference beam If the light from the two sources is in phase, constructive interference takes place, and if it is 180° out of phase, destructive interference occurs. Thus, the light detected by the sensor is a speckle pattern produced by the interference of the reference beam and the speckle pattern due to the object beam.

With the object in an undeformed state, the output from the sensor is digitised and stored on computer. This digitised data specifies a two-dimensional position description on the sensor, and the irradiance of the speckle pattern at each position. Irradiance values are stored as digitised grey levels. Grey levels from 1 to 64 or 1 to 256 are frequently used. The object is then deformed. Fig. 9.11 employs the same object beam geometry used in the previous discussion of holography. Eqn. (9.5) can again be used to determine the surface displacement magnitude. If a point on an object surface is imaged as a bright speckle spot on the sensor, this speckle will turn dark when the surface point is displaced one-quarter wavelength in the z direction. This is due to the fact that, after displacement, the optical path-length of the object beam changes by one-half wavelength and thus destructively interferes with the reference beam. The same applies to displacements given by Eqn. (9.5), where $m = \frac{1}{2}$, $\frac{3}{2}, \frac{5}{2}, \ldots$. If the speckle spot image is originally dark, after displacement it will be bright. If a point on an object surface is imaged as a bright speckle spot on the sensor, this speckle spot will be bright after displacements given by Eqn. (9.5), where $m = 0, 1, 2, 3, \ldots$. For integral values of m, if the original speckle spot image is dark, it will remain dark after displacement.

After deformation, the output from the image sensor is again digitised and stored. In order to obtain bright and dark fringes, such as those observed in holography, the grey levels from the two data sets are subtracted. The absolute value of these differences is used to produce a visual display on a video monitor. Positions where the speckle spots have the same irradiance both before and after displacement will appear dark on the video monitor, due to the subtraction process. Positions where speckle spots change from bright to dark, or vice versa, appear as bright fringes on the video monitor. Therefore, bright fringes are formed on the monitor at positions where $m = \frac{1}{2}, \frac{3}{2}, \frac{5}{2}, \ldots$, and dark fringes at positions where $m = 0, 1, 2, 3, \ldots$.

By subtracting one data set from another at a rapid rate, time-average speckle interferometry can be performed. Carrier fringes can be produced by first storing one digitised data set due to an image on the sensor in the computer memory, then rotating the object beam mirror, storing the digitised data due to a new image, and subtracting the grey levels. In this way, displacement magnitude and direction can be determined. This technique requires the same vibration isolation as holography, but the advantage is that a photographic recording medium is not required. One disadvantage is that a lens is used to image the object on the sensor, and there are therefore depth-of-field limitations. Small objects that are flat or that have large radii of curvature are excellent candidates for non-destructive testing using speckle interferometry. Equipment of this type is available (Fig. 9.12), and the technique appears to have excellent potential for solving engineering problems.

Fig. 9.12   Speckle interferometer system (Courtesy: Ealing Electro-Optics Inc.)

Chapter Ten

Laser systems in materials processing

THIS chapter deals with various types of <u>beam delivery</u> <u>systems used in industry for materials processing</u> applications. Beam delivery systems fall into three categories: <u>fixed beam</u> (moving part), <u>moving beam</u>, and hybrid (a <u>combination of moving</u> part and moving beam). For multiple-station operation using one laser, the techniques used are beam splitting, time-sharing, and multiple output beams.

The manipulation systems used include everything from simple one-axis rotary or linear translators to six- or more axis CNC and robotic systems. It would be impossible to describe all of the systems that are available (most systems are custom made), but some representative samples are shown and discussed. The choice of system depends on the type of laser used, the nature of the application(s) and the degree of flexibility required.

The material in this chapter is adapted with permission from chapters written by James Luxon for two publications; namely, 'Laser-Robot Integration for Materials Processing'. In, *Encyclopedia of Robotics*, John Wiley & Sons Inc., New York (to be published), and 'Optics for Materials Processing'. In, *The Industrial Laser Annual Handbook*, 1986 edition. PennWell Publishing Co., Tulsa, Oklahoma.

Multiple beam delivery

There are three ways of obtaining two or more beams from a single laser. These include time-sharing, beam splitting and lasers which output two or more beams directly.

In production, a laser beam is frequently time-shared between two or more workstations. This is especially useful for continuous wave (CW) CO_2 lasers when the time allotted for loading and unloading parts in the workstation is a substantial fraction of the cycle time. The best arrangement for this is a simple in-line set-up, with the laser elevated to eliminate mirrors with upward horizontal projections (Fig. 10.1). Such mirrors quickly gather dust or other debris, resulting in rapid degradation. Mirror (1) is movable, to allow the beam to go on to workstation (2). However mirror (1) is moved, whether by translation or rotation, it must be repositioned repeatedly with high accuracy in order to maintain alignment in workstation (1).

Beam splitting (Fig. 10.2) is sometimes a cost-effective way of using a higher-power laser to do two or more jobs requiring less power. (It should be mentioned that some lasers produce more than one beam and this is an alternative to beam splitting that should be considered.) The beam is on at all workstations simultaneously, unless separate power dumps (shutters) are designed into the system. In Fig. 10.2, if all three stations are to receive the same power, the reflectances of the beam splitters must be correctly designed. Beam splitter (1) should reflect 33.3% of the power incident on it, beam splitter (2) should reflect 50% (i.e. half of the remaining 66.7%), and the final mirror should reflect 100% (of the remaining 33.3%).

Another way of delivering the multiple beams to several workstations or to several locations on a given part is by means of fibre optics. At present, this is only practical for Nd-glass or Nd-YAG lasers, but it may be useful in microelectronic applications such as soldering or marking, where very little power is required to do each operation. Fibre optic beam delivery is discussed in more detail later in this chapter.

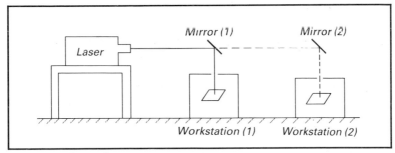

Fig. 10.1   Laser beam time-sharing

Fixed-beam systems

Many industrial dedicated systems for welding and heat treating are fixed-beam systems which utilise a parts handling system specifically designed for the application. Parts may be manually or automatically loaded and are then rotated, indexed or otherwise translated automatically. For very precise circular or flat parts, the initial focus adjustment is manual and remains fixed during processing. For sheet-metal parts or other parts or materials where part tolerances are not tight, some form of focusing axis (z-axis) control is needed. Focus control may be achieved in a variety of ways. For cutting sheet material, the focusing column may terminate in a ball-bearing device which rides on the part. The focusing head then 'floats' on this bearing fixture, which moves up and down with moderate undulations on the sheet.

For more complicated contours which cannot be followed accurately by a passive device, the surface position relative

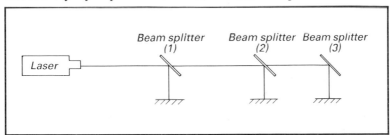

Fig. 10.2   Beam splitting. (Beam splitters are enhanced reflection-coated dielectric substrates)

to the focusing head is sensed, and the signal is fed into the computer which controls the z-axis position in order to maintain the proper focus position. The position of the cutting nozzle relative to the part is critical, and is controlled along with the focal position. Height-sensing systems employ capacitive, inductive and laser triangulation techniques. Fig. 10.3 shows a two-axis workstation employing two computer-controlled slides to form an x-y table. Such tables are capable of linear speeds of over 40cm/s with lengths of several metres. Linear and circular interpolation are used. The slides use ball screws driven by stepper or dc servo motors. The accuracy and repeatability of such systems exceed the needs of all but the most demanding applications.

Fixed-beam systems utilise both lens and off-axis parabolic mirror focusing systems. Fig. 10.4 shows an off-axis mirror-focusing accessory with a cross-jet attachment. The cross-jet attachment provides a high-speed stream of air or inert gas which runs parallel to the workpiece (perpendicular to the beam) to reduce the chances of spatter reaching the focusing mirror.

Moving-beam systems

Many systems employ beam manipulation or a combination of beam and part manipulation. This section deals with some of the systems in which the laser beam is moved from point to point in material processing.

Rectangular systems

Rectangular systems are simply x-y or x-y-z CNC systems, sometimes referred to as Cartesian systems. The beam is moved in the x-, y- and possibly the z-axis under numerical control. For relatively small and/or low-power lasers, the entire laser head may be translated in the x and y directions, whereas the z-axis motion is accomplished by control of a vertical slide or table to which the focusing assembly is attached. Fig. 10.5 illustrates a three-axis rectangular beam motion system in which the laser head is stationary.

Fig. 10.3   Two-axis CNC system

In this configuration, mirror (2) is translated to give the *x* motion, both mirrors are moved together to provide the *y* motion, and the focusing assembly is moved vertically to provide the *z*-axis motion. In some hybrid systems, the beam

Fig. 10.4 Parabolic mirror focusing unit

is moved in the x direction by moving mirror (2), and the part is moved in the y direction. The z-axis is often automatically controlled, mechanically or electronically, for cutting sheet material.

Fig. 10.6 shows a two-axis sheet-cutting system which combines beam motion and sheet motion with an adaptive z-axis. For three-dimensional parts, pure rectangular three-axis motion is not sufficient. Many different types of system are available. Some utilise a combination of rectangular part manipulation and angular beam manipulation, while others utilise all beam manipulation. Redundant axes may be used; for example, large translational motion in one direction may be achieved by part movement, with the smaller movements being accomplished by beam manipulation. Fig. 10.7 illustrates a beam positioner with two angular degrees of freedom. Translation of the entire fixture vertically gives the z translation.

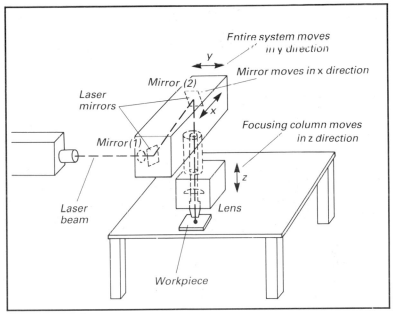

Fig. 10.5   Rectangular beam motion system

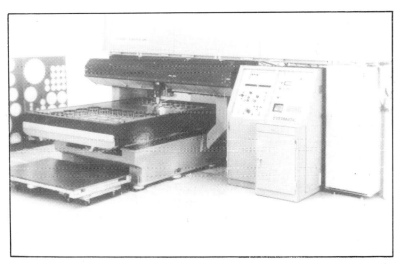

Fig. 10.6   Two-axis sheet-cutting system (Courtesy: Coherent General Inc.)

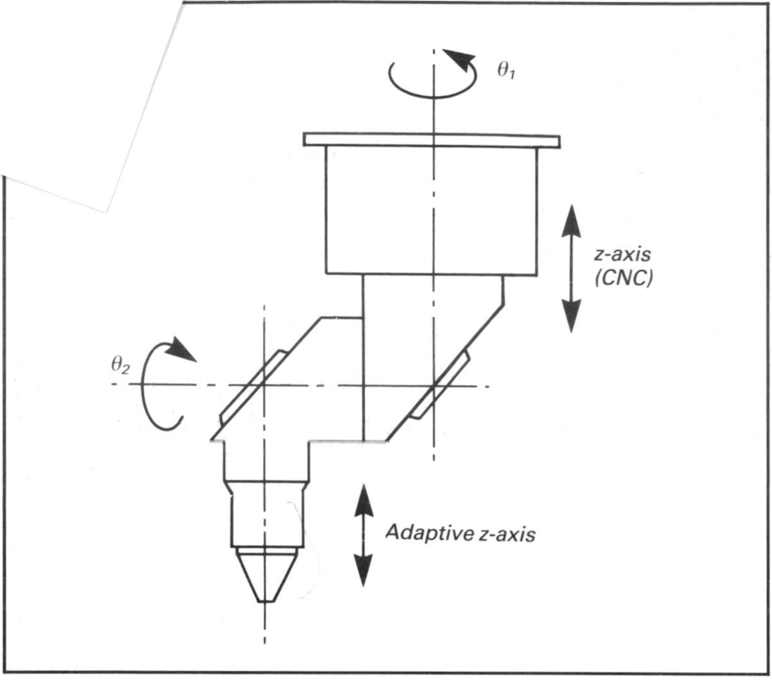

Fig. 10.7   Focusing fixture with two rotational degrees of freedom,
adaptive focus adjustment, and z-axis numerical control

Laser-robot integration

The trend towards the integration of lasers and robots is a
natural consequence of the flexibility of robots and the
adaptability of lasers to flexible systems. The first step in the
development of laser-robot systems was the integration of
existing lasers with existing robots. This included both
rectangular coordinate (gantry) robots and articulated-arm
robots (Figs. 10.8 and 10.9).

A wide variety of gantry-type systems have been developed for use with high-power CO_2 lasers of about 500W –
25kW power output. In some of the lower power systems,
where the laser heads are relatively small and light, the laser
head is mounted on the gantry and is translated in the x and

Fig. 10.8 Gantry robot (Courtesy: ILM Inc.)

Fig. 10.9 Through-the-arm laser robot (Courtesy: ILM Inc.)

y directions. The *z*-axis telescopes may terminate in a two-
or three-axis angular beam manipulation device, and possibly an adaptive *z*-axis control. The advantages of this type
of design are that it eliminates two bending mirrors and
limits the beam path length variation to the *z*-axis motion,
which is usually less than one metre. The disadvantage is
that the laser head, which even for a 500W laser may weigh
90kg must be moved in the *x* and *y* directions, thus possibly
limiting acceleration, speed and accuracy.

Another form of gantry adaptation is to mount the laser
head in a stationary position and to use the robot to move a
mirror system in the *x* and *y* directions. The advantage of
this method, obviously, is the reduced weight the robot has
to carry. However, the disadvantage is that the beam path
length varies by up to several metres for large systems.

In applications where rolled sheet or other long parts are
to be cut and/or welded, one rectangular axis may be
associated with the part in order to move it through the
beam work area as well as to provide one axis of motion.
This technique has the advantage of permitting the system to
handle rolls or parts of virtually any length. It also reduces
beam path length variation. For large parts, such as car
bodies, automated guided vehicles (AGVs) are used.

The accuracy, repeatability and speed of many gantry
systems are excellent, and they make good use of the laser's
capabilities when integrated. Generally, rectangular systems
are simpler to program, although teach modes and optical
follower systems are employed in some applications. Existing gantry robots have been integrated with lasers for
research purposes with considerable success. A number of
custom-made laser-gantry robot systems have also been
built, particularly for the automotive industry.

Articulated-arm robots potentially afford the greatest
amount of flexibility when integrated with a laser. Considerable effort has been, and continues to be, spent on the
integration of Nd-YAG, Nd-glass and CO_2 lasers with
articulated-arm robots.

The predominant method of beam delivery for the

integration of solid lasers with robots is the fibre optic light guide. Several groups are doing research into this technique and fibre optic beam delivery systems (Fig. 10.10) are available commercially.

The fibres used for this purpose are fused silica step-index fibres of between about 0.5mm and 1mm in diameter. The

Fig. 10.10   Fibre optic beam delivery system

losses in these fibres for the 1.06μm radiation of Nd-YAG or Nd-glass lasers are about 6dB/km. For the lengths of fibres likely to be used in industry, this translates into a negligible loss. The ends of the fibres are not anti-reflection-coated, so a loss of 8–10% can be expected due to reflection. Other losses will be incurred due to reflection from focusing and collimating lenses. Overall losses in the order of 15% are reasonable.

The mechanism for beam propagation in a step-index fibre is total internal reflection at an interface between the fibre core and a clad layer of slightly lower refractive index. The fibre has a numerical aperture (NA) which is given by:

$$NA = \sin\theta = (n_{core}^2 - n_{clad}^2)^{1/2} \qquad (10.1)$$

where n is the refractive index and θ is the half-angle of the entrance cone for light entering or leaving the fibre. Light power which corresponds to rays striking the end of the fibre at an angle greater than θ will be quickly attenuated in the fibre. The entrance angle is about 20°. Hence, the laser beam must be focused on the end of the fibre to a spot size slightly less than the fibre diameter, with the proper convergence angle. The beam leaving the fibre diverges and must be collimated prior to final focusing. Systems now available commercially include the fibre, optics, and reinforced cable for mechanical protection of the optical fibre.

Care must be taken in the use of such cables to ensure that the cable is not bent too sharply, although bend radii of 20cm can be tolerated for a 1mm fibre cable. The concern here is not breakage, but greatly increased absorption. The fibre will not tolerate any twisting and bending simultaneously, otherwise losses will be dramatically increased.

One end of the fibre optic cable is attached to the laser, and the other is placed in an appropriate gripper at the end of the robot's wrist. The same robot could be used for pick-up and placement of parts, if the gripper is suitably designed to handle the parts as well as the end of the fibre optic cable. Power levels of up to 400W have been propagated through fibres over 25m long. However, the authors are not aware of any reports of extended use of fibres at high average power level.

Most existing small robots capable of handling the lightweight cables cannot take full advantage of the capability of the fibre optic beam delivery system. Such robots vibrate, causing excessive kerf width and lower cutting speeds, and do not maintain constant speed while going round sharp corners. Improved laser-robot systems will provide more stable platforms for the laser beam, as well as either speed monitoring to automatically modulate the laser power, or true constant speed motion of the end-effector.

The first generation of CO_2 lasers integrated with articulated-arm robots employ a beam delivery system made out of mirrors and conduit-connected by joints or 'knuckles' which provide for flexibility, in much the same way as the robot arm gives the robot its ability to reach and rotate. One system uses a telescoping conduit, which reduces the number of knuckles required for extended reaching. The telescoping type causes some variation in the beam path length, whereas the non-telescoping systems hold the path length constant, but require several more mirrors in the system in order to do so. A telescoping system may have as few as four mirrors, compared with as many as 11 in non-telescoping types. Each mirror can be expected to absorb about 1% of the power when new, rising to 3% or more as it ages.

Figs. 10.11(b) and (c) show two types of mirror-conduit beam delivery systems, and Fig. 10.11(a) two knuckles, each containing three mirrors.

When coupled with an appropriate two- or three-axis rotational beam motion end-effector placed at the end of the robot's wrist, this type of system provides a high degree of flexibility for three-dimensional material removal or welding. However, these beam delivery systems do not have the same working envelope as the robot and can be easily damaged. It is advisable for the robot to be retrofitted with limit switches or other sensing devices, which will stop it if it

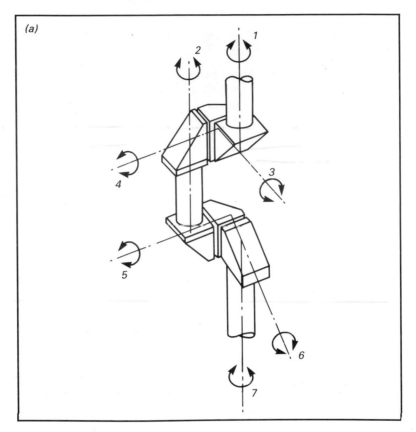

(a)

approaches the limit of the working envelope of the beam delivery system. If the robot makes a sudden rapid motion, a knuckle may bend in an unexpected direction, or if the robot causes either the end-effector or the conduit to collide with something, damage is likely to occur. Extremely precise alignment is required in these systems and they must be returned to the manufacturer for repair if such damage occurs as a result of misuse.

A purge of clean, dry air or inert gas must always be maintained in the conduit to prevent contamination of the mirror surfaces, which rapidly increases absorption losses. Depending on the power density on the mirrors, they may have to be watercooled. Cooling should be seriously

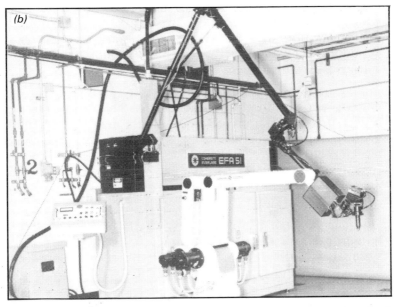

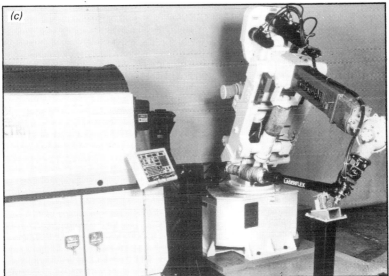

Fig. 10.11 CO_2 laser-robot beam delivery systems: (a) knuckles used in systems, (b) and (c) two types of mirror-conduit beam delivery systems (Photos courtesy: Coherent General Inc. and Spectra Physics, Industrial Laser Division)

considered for systems handling more than 1000W. Systems which can handle up to 5kW are available commercially. Losses in new systems vary from about 15% to 30%, depending on various factors, such as the number of mirrors. These losses will increase with time.

The next generation of CO_2 laser-robot systems will employ two different approaches. Systems using the new breed of compact transverse-flow lasers may have the laser head mounted on the robot shoulder, with the beam directed either through the arm or through a conduit which runs parallel to the arm. This overcomes the problems associated with different working envelopes for the beam delivery system and the robot.

For larger higher-power lasers, where the laser head must be mounted off the robot, through-the-arm beam delivery systems will still be employed for the reason mentioned above. These systems will be highly stable, with great accuracy and repeatability. Accurate speed control will also be necessary in order to take full advantage of the laser's capabilities.

Bibliography/
Further Reading

Books

Hitz, C. B., 1985. *Understanding Laser Technology.* PennWell Books, Tulsa, Oklahoma. **[1,2]**

Luxon, J. T. and Parker, D. E., 1985. *Industrial Lasers and Their Applications.* Prentice-Hall, Engelwood Cliffs, New Jersey. **[1–9]**

Svelto, O. and Hanna, D. C., 1982. *Principles of Lasers* (2nd edn.). Plenum Press, New York. **[1,2]**

Young, M., 1986. *Optics and Lasers* (3rd edn.). Springer-Verlag, New York. **[1–3]**

Duley, W. W., 1976. CO_2 *Lasers – Effects and Applications.* Academic Press, New York. **[3–7]**

Belforte, D. and Levitt, M. (Eds.), 1986. *The Industrial Laser Annual Handbook.* PennWell Books, Tulsa, Oklahoma. **[5–7,10]**

Bass, M. (Ed.), 1983. *Laser Materials Processing.* North-Holland, New York. **[5–7,10]**

Marshall, G. (Ed.), 1985. *Laser Beam Scanning.* Marcel Dekker, New York. **[8]**

Vest, C., 1979. *Holographic Interferometry.* John Wiley, New York. **[9]**

Ballard, D. and Brown, C., 1982. *Computer Vision.* Prentice-Hall, Engelwood Cliffs, New Jersey. **[8,9]**

Buyers' Guides

Laser Focus Buyers Guide. PennWell Publishing Co., Littleton, Massachusetts.

Lasers and Applications Buying Guide. High Tech, Torrance, California.

Photonics Spectra Buyers Guide. Laurin Publishing Co., Pittsfield, Massachusetts.

Industrial Lasers – Productivity Equipment Series. Society of Manufacturing Engineers, Dearborn, Michigan.

Trade Journals

Laser Focus. PennWell Publishing Co., Littleton, Massachusetts.

Lasers and Applications. High Tech, Torrance, California.

Photonics Spectra. Laurin Publishing Co., Pittsfield, Massachusetts.

Index